The Cloud Computing Book

The Cloud Computing Book

The Future Of Computing Explained

DOUGLAS E. COMER

Department of Computer Sciences
Purdue University
West Lafayette, IN 47907

CRC Press
Taylor & Francis Group
Boca Raton London New York

CRC Press is an imprint of the
Taylor & Francis Group, an **informa** business

First edition published 2021
by CRC Press
6000 Broken Sound Parkway NW, Suite 300, Boca Raton, FL 33487-2742

and by CRC Press
4 Park Square, Milton Park, Abingdon, Oxon, OX14 4RN

Visit the Taylor & Francis Web site at
http://www.taylorandfrancis.com

and the CRC Press Web site at
http://www.crcpress.com

Library of Congress Cataloging-in-Publication Data

Names: Comer, Douglas, author.
Title: The cloud computing book : the future of computing explained / Douglas Comer.
Description: First edition. | Boca Raton : CRC Press, 2021. | Includes bibliographical references and index.
Identifiers: LCCN 2020052980 | ISBN 9780367706807 (hbk) | ISBN 9781003147503 (ebk)
Subjects: LCSH: Cloud computing.
Classification: LCC QA76.585 .C636 2021 | DDC 004.67/82--dc23
LC record available at https://lccn.loc.gov/202005298

ISBN: 978-0-367-70680-7 (hbk)
ISBN: 978-1-003-14750-3 (ebk)

Typeset in Times font
by KnowledgeWorks Global Ltd.

*To Chris, who takes
me to cloud nine*

Contents

Chapter 3   Types Of Clouds And Cloud Providers   25

PART II Cloud Infrastructure And Virtualization   33

Chapter 4   Data Center Infrastructure And Equipment   37

Chapter 5   Virtual Machines   **55**

Chapter 6   Containers   **71**

Chapter 7   Virtual Networks   **87**

Chapter 8   Virtual Storage   99

PART III Automation And Orchestration   109

Chapter 9   Automation   113

Chapter 12 Microservices

Chapter 13 Controller-Based Management Software

Chapter 14 Serverless Computing And Event Processing

Chapter 18   Controlling The Complexity Of Cloud-Native Systems   247

Index   265

Preface

In late 2020, a headline in a tech industry newsletter announced:

Cloud Is Changing Everything

Although it may seem like hype, the headline captures a fundamental change in computing. Organizations, both large and small, are migrating computing and data storage to the cloud, which has grown an entirely new ecosystem. Services ranging from Internet search, AI, and big data analysis to individual document processing use cloud facilities.

Curricula in Computer Science and Computer Engineering must shift to prepare students for cloud computing. Instead of simply learning how to design and use software and hardware systems for the traditional mobile and desktop world, courses must prepare students to design and build hardware and software systems for use in a cloud data center. Instead of merely building a conventional computer program and relying on faster hardware to increase performance, cloud employs a parallel approach that requires one to deploy and manage multiple copies of each software system.

What should one learn that will help them navigate constant change? A colleague quipped that courses on cloud will quickly sink into teaching technology because commercial technologies dominate the cloud industry. Indeed, dozens of technologies exist, and new technologies seem to appear every few months, forcing constant churn. The answer is that colleges and universities should teach basic concepts and principles that will remain valid despite the ever-changing commercial world. For example, to master new orchestration technologies, one needs to understand the big picture of why orchestration is needed and how orchestration systems work. Similarly, to make sense of Virtual Machine and Container technologies, one needs to understand the fundamentals of each. To understand why a cloud provider may scatter a tenant's virtual servers across a set of physical servers, one must understand the underlying infrastructure and surprising topics such as power and cooling.

This text provides a broad overview of cloud computing, covering all aspects from basic data center facilities to the ways cloud-native software differs from traditional software. Instead of describing the services offered by a particular public cloud company or attempting to cover the wide range of third-party offerings, the text concentrates on general concepts that span many providers and services. To help keep the discussion focused on reality, the text uses concrete examples of technologies that currently dominate the industry, including Docker Containers and Kubernetes Orchestration technology. However, rather than attempting to present all details, the text gives examples that show the essence of each.

Designed as both a textbook and professional reference, the text is suitable for a one-semester course in computer science and engineering and for professional software engineers and IT staff who need to understand cloud. The material has been divided into five parts. The first part describes the motivation, advantages, and growth of cloud computing. The second part describes cloud infrastructure and virtualization, including virtual computing, networking, and storage mechanisms. The third part describes high-level automation and orchestration systems that manage the virtualized infrastructure. The fourth part describes cloud software, including the programming paradigms used and how cloud software deployments scale to large numbers of users. The final part describes remaining topics, including the concept of edge computing and its relationship to the Industrial Internet of Things, security problems that arise in a cloud environment, and approaches that help designers control the complexity of cloud deployments.

One cannot appreciate cloud computing without hands-on experience. Fortunately, major cloud providers offer "free tier" services that allow individuals to try deploying cloud services at no cost; some providers also offer special educational accounts and curricula materials†. Although free accounts have limited computational and storage resources, they allow one to learn about the technologies and interfaces available from various cloud providers. I strongly encourage everyone who wants to understand cloud to obtain one or more free accounts and deploy a VM (e.g., a web server), use Docker to deploy a container, and (for more advanced readers) use Kubernetes to orchestrate replicated containers. Examples of free tier accounts include:

Amazon Web Services free tier	**https://aws.amazon.com/free/**
Google Cloud Platform free tier	**https://cloud.google.com/free/**
Azure Cloud free tier	**https://azure.microsoft.com/free/**
IBM Cloud free tier	**https://www.ibm.com/cloud/free**
Oracle Cloud	**https://www.oracle.com/cloud/free/**

I thank many individuals and groups who contributed to the book. Daniel J. Krolopp, Adib Rastegarnia, Paul Schmitt, Phil Van Every, and John Lin proofread chapters and provided suggestions about the content. Adib developed the TLA+ example in Chapter 18. Ted Turner provided insights on trends in the industry. Nick Lippis from the Open Network User's Group graciously invited me to participate in meetings, where I learned how Fortune 100 firms are moving to a hybrid multi-cloud infrastructure. Aryo Kresnadi helped me understand the scale of Fed Ex facilities and their use of cloud. Ernest Leffler introduced me to how the banking and financial services industries are making use of cloud. Scott Comer, Sharon Comer, and Mark Kunschke provided support and suggested cover designs.

Finally, I thank my wife, Christine, for her patient and careful editing and valuable suggestions that improved and polished the text.

Douglas E. Comer

About The Author

Douglas E. Comer is a Distinguished Professor at Purdue University in the departments of Computer Science and Electrical and Computer Engineering (courtesy). He has an extensive background in computer systems, and has worked with both hardware and software.

One of the researchers who contributed to the Internet as it was being formed in the late 1970s and 1980s, he has served as a member of the Internet Architecture Board, the group responsible for guiding the Internet's development. Prof. Comer is an internationally recognized expert on computer networking, the TCP/IP protocols, and the Internet.

In addition to research articles, he has written a series of widely acclaimed textbooks on the topics of computer architecture, operating system design, computer networking, and the Internet. Prof. Comer's books have been translated into many languages, and are used in industry as well as Computer Science, Engineering, and Business departments around the world. He continues to consult and lecture at universities, industries, and conferences.

While on leave from Purdue, Prof. Comer served as the inaugural VP of Research at Cisco Systems. For twenty years, Prof. Comer served as the editor-in-chief of the journal *Software —Practice and Experience* . He is a Fellow of the Association for Computing Machinery (ACM), a Fellow of the Purdue Teaching Academy, and a recipient of numerous awards, including a USENIX Lifetime Achievement Award. Prof. Comer is a member of the Internet Hall of Fame.

Additional information can be found at:

www.cs.purdue.edu/people/comer

and information about Comer's books can be found at:

www.comerbooks.com

Part I

Why Cloud?

The Motivation, Advantages, And Growth Of Cloud Computing

Chapter Contents

1

The Motivations For Cloud

1.1 Cloud Computing Everywhere

In a short time, cloud computing has become significant. Consider the following examples of situations that involve cloud computing.

- A startup leases cloud facilities for its web site; the company can pay for additional facilities as web traffic grows.

- An individual uses a smart phone to check *Internet of Things* (*IoT*) devices in their residence.

- An enterprise company leases facilities and software for business functions, such as payroll, accounting, and billing.

- Students working on a team project use a browser to edit a shared document.

- A patient wears a medical device that periodically uploads readings for analysis; their doctor is alerted if a medical problem is detected.

- A seasonal company leases computing facilities during four peak months each year; the company doesn't pay for facilities at other times.

- A teenager logs into a social media site and uploads photos.

- A retail company leases facilities at the end of each fiscal year to run data analytics software that analyzes sales for the year.

- An individual uses a streaming service to watch a movie; a copy of the movie is kept in a facility near the family's residence.

- The recipient of a package uses a tracking number to learn about the current location of the package and the expected delivery time.

Most enterprises —not just high tech firms and social media companies —are moving to the cloud. In the early 2000s, business functions, such as payroll, accounting, billing, and supply chain management were implemented with local facilities operated by an organization's IT staff. Now, such functions are being migrated to cloud computing. Enterprises that do decide to retain some local computing are shifting from a paradigm of having services run on individual computers located in various departments throughout the organization to a paradigm where the facilities are consolidated into a local cloud data center.

In addition to providing communication to consumer IoT devices, the cloud provides processing. For example, battery-operated IoT sensors are used to monitor civil infrastructure, such as bridges. A set of sensors on a bridge periodically measures vibrations or stress and uploads the data to the cloud. Software running in the cloud data center combines measurements from dozens of sensors and assesses the safety of the bridge. The amount of processing needed for such computations far exceeds the capability of battery-operated sensors.

1.2 A Facility For Flexible Computing

Many of the examples illustrate a key aspect of cloud computing: flexibility to accommodate both incremental growth and cyclic demand. A cloud offers flexible computing facilities (servers and software), storage facilities, and communication facilities (Internet connections).

Incremental growth. The startup scenario shows why incremental growth is important. A small startup can begin by leasing minimal cloud facilities (e.g., enough to support a basic web site), and then increase its lease as the business grows. Similarly, the startup can begin by leasing minimal software (e.g., basic web and payment processing software), and then add database and accounting software later. The cloud provider will be able to satisfy computing needs even if the startup grows into a substantial enterprise business.

Cyclic demand. Even if a company does not engage in seasonal business, demand for computing changes throughout the year. Reports may be generated at the end of each month or each quarter as well as at the end of the year. Sales activity and order processing often spike at the end of the month as sales staff work to meet their monthly quotas. Cloud allows companies to lease additional facilities to accommodate such demand.

Although the ability to lease resources as needed is attractive, the most significant aspect arises from the pricing model: a cloud provider only charges the customer for the facilities they actually use.

> *Cloud computing allows each customer to increase or decrease their use of cloud facilities at any time; a customer only pays for the facilities they use.*

1.3 The Start Of Cloud: The Power Wall And Multiple Cores

What has motivated the shift to cloud computing? This chapter considers how cloud arose. The next chapters provide an overview of cloud, and describe the rise of cloud providers. Later chapters explore cloud infrastructure and technologies.

Two intertwined factors contributed to the start of the cloud paradigm.

- Technological: limits on speed forced a move to parallelism

- Economic: changes in technology changed IT costs

We begin by looking at the technological factor, and then examine how the new technology changed the cost of acquiring and running equipment and software.

Throughout the 1980s and early 1990s, chip manufacturers produced a series of processors that had more functionality and higher speed than the previous model. As a consequence, individual computers became more powerful each year while costs remained approximately the same. The availability of powerful, low-cost computational facilities encouraged both individuals and organizations to expand their use of computers. At any time, if the processing power of a computer became insufficient for the workload, a computer could easily be upgraded to a newer, more powerful model. In particular, to offer internal and external services, such as the World Wide Web, an organization could run the software on a powerful computer known as a *server*. When demand for a particular service became high, the organization could replace the server being used with a model that had a faster processor and more memory.

By the late 1990s, the chip industry faced a serious limitation. Moore's Law — the prediction that the number of transistors would double every eighteen months — meant that the size of a given transistor was shrinking. More transistors were squeezed together on a chip each year. Each transistor consumes a small amount of power and emits a small amount of heat. When billions of transistors are squeezed together in a small space, the temperature can climb high. More important, the amount of power consumed —and therefore, the temperature of the chip —is proportional to the square of the clock speed. Consequently, even a small increase in clock speed increases the temperature significantly. Manufacturers eventually reached a critical point, and processor speeds could not be increased beyond a few Gigahertz without generating so much heat that the chip would burn out. Industry uses the term *power wall* to characterize the limit on processor speed.

How can additional computational power be achieved without increasing the speed of a processor? The answer lies in parallelism —using multiple processors that each operate at a speed below the power wall instead of one processor that operates at a super high speed. To handle the situation, chip manufacturers devised chips that contain multiple copies of a processor, known as *cores*. Each core consists of a complete processor that operates at a safe speed, and software must devise a way to use multiple cores to perform computation.

Industry uses the term *multicore processor* to describe a chip with multiple processors. A *dual core* chip contains two processors, a *quad core* chip contains four, and so on. Multicore processors form one of the fundamental building blocks for cloud computing. Unlike the processors used in consumer products, however, the multicore processors used in cloud systems have many cores (e.g., 64 or 128). A later chapter examines virtualization software and explains how cloud systems use computers with multiple cores.

1.4 From Multiple Cores To Multiple Machines

Although they offer increased processing power on a chip, multiple cores do not solve the problem of arbitrary scale. The cores on a chip all share underlying memory and I/O access. Unfortunately, as the number of cores increases, I/O and memory accesses become a bottleneck.

How can more powerful computers be constructed? The science research community was among the first groups to explore a design that provided the basis for cloud. As scientific instruments, such as colliders and space telescopes, moved to digital technologies, the amount of data grew beyond the capabilities of even the most powerful supercomputers. Furthermore, a supercomputer was an extremely expensive machine; few universities and laboratories could afford to purchase multiple supercomputers. However, personal computers had become commodity items as reflected by their low price. Despite the drop in price, personal computers had also become more powerful. Scientists wondered if instead of using expensive supercomputers, a new form of supercomputing could be achieved by interconnecting a large set of inexpensive personal computers. The resulting configuration, which became known as a *cluster architecture*, has a key advantage: processing power can be increased incrementally by adding additional inexpensive commodity computers.

Using multiple computers for scientific computations poses a software challenge: a calculation must be divided into pieces so that each piece can be handled by one of the smaller computers in the cluster. Although the approach does not work well in some cases, the science community found ways to use a cluster for many of its important computational problems. Thus, the cluster architecture became accepted as the best way to build affordable, incrementally expandable supercomputers.

1.5 From Clusters To Web Sites And Load Balancing

As the World Wide Web grew in popularity in the 1990s, the traffic to each web site increased. As in the science community, the limitation on the speed of a given processor presented a challenge to the staff responsible for running web sites, and they also considered how to use multiple personal computers to solve the problem.

Web sites and scientific computing systems differ in a fundamental way. Supercomputer clusters intended for scientific calculations are designed so that small computers can work together on one computation at a time. In contrast, a web site must be designed to process many independent requests simultaneously. For example, consider a retail web site. One user may be looking at tools while another shops for clothing, and so on, with little overlap among the web pages they visit.

The question arose: how can a web site scale to accommodate thousands of users? Part of the answer came from a technology that has become a fundamental component in cloud computing: a *load balancer*. Typically implemented as a special-purpose hardware device, a load balancer divides incoming traffic among a set of computers that run servers. Figure 1.1 illustrates the idea. As the figure shows, a server may need to access a company database (e.g., to validate a customer's account during checkout).

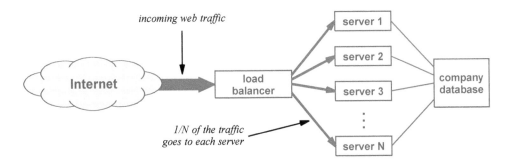

Figure 1.1 Illustration of a load balancer used to divide incoming network traffic among a set of computers.

Load balancing technology ensures that all communication from a given customer goes to the same server. The scheme has a key advantage: successive requests from a customer go back to the server that handled earlier requests, making it possible for the server to retain information and use it for a later request. From the web site owner's point of view, if the site has N servers, load balancing means that each server handles approximately $1/N$ of the customers. At any time, a site could be expanded merely by adding additional servers; the load balancer automatically includes the new servers when dividing traffic.

1.6 Racks Of Server Computers

As demand increased for facilities composed of many smaller computers, computer vendors responded by changing the shape of computers to make it easier to store many computers in a small space. Instead of large enclosures that had significant amounts of empty space inside, designers focused on finding ways to reduce the size. Furthermore, they redesigned the enclosures to fit into tall metal equipment cabinets called *racks*. Before computers, racks had been used for telephone equipment, and were widely available. A full-size rack is approximately six and one-half feet tall, two feet wide, and three and one-half feet deep. The rack contains metal mounting bars called *rails* to which equipment can be attached.

Server computers can be stacked vertically in a rack. A full rack contains forty-two *Units* of space, where a Unit is 1.752 inches. A server is designed to be one unit tall, written *1U*. In principle, one could stack forty-two 1U servers in each rack, as Figure 1.2 illustrates.

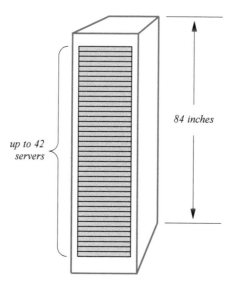

Figure 1.2 Illustration of a rack that holds multiple servers.

In practice, additional constraints usually mean that a rack is not full of servers. For example, at least one slot in a rack (usually the top of the rack) contains a network switch used to provide network connections for the servers in the rack. In addition, some slots may be left empty to allow air flow to keep the servers from overheating.

1.7 The Economic Motivation For A Centralized Data Center

The availability of low-cost servers and the ability to collect multiple servers into a rack may seem insignificant. From the point of view of IT management, however, collecting servers into a small place has an important advantage: lower cost. There are two aspects:

- Operating expenses (opex): lower recurring cost
- Capital expenses (capex): lower equipment cost

Lower operating expenses. To understand how placing servers in racks can reduce operating expenses, recall that the low cost of computer hardware made it easy for each group in a large organization to acquire its own computing facilities. While hardware costs were falling, the recurring cost of IT staff increased because demand for trained IT professionals caused salaries to rise. Each group hired staff, and because technology changes rapidly, each group had to pay for training to keep their staff up-to-date.

Ironically, the availability of low-cost personal computers, which everyone assumed would lower overall expenses, created a situation in which an organization faced much higher costs because each department purchased many computers and then hired their own staff to manage them. By the 2000s, many organizations became concerned about rising IT costs. An executive at one company quipped,

> Cheap computers have turned into a major expense.

Organizations faced an IT staffing dilemma. On the one hand, allowing every department to hire and train its own IT staff to maintain skills results in unnecessary duplication. Some of the skills are only needed infrequently when problems arise. On the other hand, although it helps share expertise and reduce overall training costs, consolidating IT staff into a central department has the disadvantage that when a problem occurs, a staff member must travel to the department that owns the server to assess the problem and work on the equipment. For an organization with a large campus, such travel may require hiring additional staff.

The availability of high-speed computer networks allows an organization to optimize costs by consolidating server equipment into a single physical location. Instead of locating a server in each department, the organization places all servers in racks in a central facility, and hires a small, centralized staff to maintain the servers. Employees in departments can access the servers over a network. Such a centralized facility has become known as a *data center*†.

Lower capital expenses. The data center approach has an advantage of reducing overall equipment cost. If server computers are distributed throughout an organization, each individual department must choose when to upgrade their server. In addition, each department will consider the cost of a new server as well as the features and performance they need. As a result, a department will usually order one server at a time, and the configuration ordered by one department is likely to differ from the configuration ordered by another department.

If an organization consolidates servers into a data center, the organization can choose a uniform configuration for all servers. Furthermore, the organization can upgrade many servers at once (e.g., upgrade one-third of all servers every year). Consequently, when an upgrade occurs, the organization will purchase dozens or hundreds of servers at the same time, making it possible to negotiate a quantity discount.

We will learn that cloud providers exploit and extend the economic benefits described above. Not only do their data centers benefit from lower capex and lower opex, they also employ technologies that increase or decrease available resources dynamically, allowing them to achieve better overall utilization.

---

†Although the term *computing center* may seem more appropriate than *data center*, industry had used the term *computing center* decades earlier to describe a centralized mainframe facility; using a new name helped avoid the impression of moving backward.

1.8 Origin Of The Term "In The Cloud"

As Figure 1.1† illustrates, we use a cloud to depict the Internet in diagrams. The cloud represents a set of networks and the equipment that interconnects the networks. Saying a data center is "in the cloud" is technically inaccurate because the servers in a data center are not part of the Internet itself. Instead, servers are merely computers that attach to the Internet, and should be depicted outside the cloud, with network connections leading to the cloud.

Why did the industry start using the terms *cloud computing* and say that the computing is *in the cloud* if it is not? Early data centers that supplied large-scale web services needed high-capacity connections to major Internet backbone networks. Placing a data center near an Internet *peering point* where major Internet backbones interconnect minimizes cost. Some providers actually placed their data centers on the same floor in the same building as an Internet peering point. Because such a data center was located with networking equipment rather than in a separate building, engineers said the data center was *in the cloud*, and the terminology caught on. The point is:

> *Although technically inaccurate, the phrase* in the cloud *arose because early data centers were located close to networking equipment at Internet peering points rather than in separate buildings.*

1.9 Centralization Once Again

Surprisingly, after many decades of increasing decentralization, the move to data centers reverses the trend and moves back toward a more centralized model. The next chapter considers how public cloud providers extend centralization further by creating data centers with servers for multiple organizations. We can summarize:

> *For decades, the low cost of computers encouraged decentralization. The power wall and cost of IT staffing favor a return to a centralized model that consolidates computing facilities into data centers.*

---

†Figure 1.1 can be found on page 9.

Chapter Contents

2

Elastic Computing And Its Advantages

2.1 Introduction

The previous chapter describes the power wall that forced the computing industry to move to the use of multiple processors and multiple computers. The chapter also discusses how collecting servers into a centralized facility can reduce both the cost of equipment and the recurring cost of IT staff.

This chapter explains how software enables a set of servers to be used in an entirely new way. It describes the key concept of elastic computing and explains how virtualization enables cloud owners to scale services. The chapter also describes business models that arose in the cloud industry. The next chapter describes how public cloud companies arose that use elastic computing. Later chapters explain the underlying infra structure and technologies that enable elastic computing as well as software systems used to manage deployments in a cloud.

2.2 Multi-Tenant Clouds

The previous chapter points out that consolidating servers into a single data center has economic advantages for an organization because the organization can negotiate quantity discounts on the purchase of equipment and the organization spends less on the recurring cost of maintaining and training IT staff. Cloud providers extend the advantages to a larger scale. In particular, instead of handling computing for one organization, a cloud provider builds a data center (or multiple data centers) that can handle computing for many customers.

We use the term *multi-tenant* to refer to a data center that serves customers from multiple organizations. We will see that the technologies used in cloud systems are designed to support multi-tenant computing and keep the data of each customer safe. Interestingly, the idea of multi-tenant clouds applies to groups within a single organization. For example, the finance department may want to keep all data completely separate from the rest of the organization, or a business unit may choose to keep its records separate from other business units.

2.3 The Concept Of Elastic Computing

As the examples in Chapter 1 show, a fundamental aspect of cloud computing centers on the ability of a customer to lease servers and only pay for the number of servers they need. A customer can choose to lease a few servers or many. More important, a customer can change the allocation dynamically, adding servers during peak times and decreasing the number of servers during times they are not needed. We use the term *elastic computing* to describe such a dynamic service.

How can a cloud owner increase and decrease the number of servers allocated to a given customer? One early approach allocated a set of physical servers. A provider restricted a customer to a few sizes. For example, a customer could lease a full rack of servers, a half rack, or a quarter rack, or multiples of the sizes. However, such an approach is relatively inflexible because it requires a provider to dedicate a set of physical resources to each customer.

2.4 Using Virtualized Servers For Rapid Change

A technology emerged that allows a cloud provider to offer fast, flexible allocation at low cost. Surprisingly, the technology does not involve allocating sets of physical servers. Instead, a cloud owner runs software on each physical server that allows the cloud owner to create a set of *virtualized servers*. Later, we will learn more about virtualization technologies, such as *Virtual Machines* and *Containers*; for now, it is sufficient to understand three important properties of virtualized servers:

- Rapid creation and removal
- Physical sharing
- Logical isolation

Rapid creation and removal. Virtualized servers are managed entirely by software. Management software can create or remove a new virtualized server at any time, without changing or rebooting physical servers.

Physical sharing. Because it consists of a software artifact, a virtualized server is similar to a computer program. Each virtualized server must run on a physical server, and multiple virtualized servers can run on a given physical server concurrently.

Logical isolation. Although multiple virtualized servers can run on a single physical server at the same time, each virtualized server is completely isolated from the others. Thus, the data and computations performed by one virtualized server cannot be observed or affected by another.

2.5 How Virtualized Servers Aid Providers

From a cloud provider's point of view, the ability to virtualize servers provides the basis for elastic computing and makes cloud computing economically viable. A cloud provider only needs to use computer software to increase or decrease the number of servers a customer is leasing. Furthermore, the use of virtualized servers allows a provider to accommodate the changing needs of many customers. Even with the overhead of checking a customer's account and configuring the newly-created virtualized server, creation takes milliseconds, not minutes or hours. More important, because the provider does not need to reconfigure or reboot physical servers, a provider can accommodate the creation of thousands of virtualized servers as needed.

Because virtualization technologies guarantee isolation, a cloud provider can place virtualized servers on physical servers without regard to the owner, the apps they will run, or the data they will handle. In particular, isolation allows a provider to mix virtualized servers from multiple customers on the same physical server without any interference and without any chance of data from one customer's virtualized server "leaking" to another customer.

The ability to run arbitrary virtualized servers together on a given physical server means that when a provider needs to choose where to run a virtualized server, the choice can be made in a way that provides the best benefit to the provider, without regard to the owner of the new virtualized server or the owners of the other virtualized servers that are running. In particular, a provider tries to avoid placing too many virtualized servers on the same physical server because each virtualized server will receive less processing power, causing customer complaints. To avoid such problems, a provider can use virtualized server placement to balance the load across all physical servers in the data center†. That is, when creating a virtualized server, a provider can choose a physical server that is lightly loaded. Doing so does not guarantee an equal load, but it definitely avoids overloading one physical server while others are idle.

To summarize:

> *Using virtualized servers provides invaluable advantages for cloud providers, including the ability to scale the service and the ability to balance the load and avoid overloading a physical server while other physical servers remain idle.*

---

†This chapter describes balancing load when creating a virtualized server; Chapter 7 describes how a virtualized server can be moved from one physical server to another while it is running.

2.6 How Virtualized Servers Help A Customer

To a customer, a virtualized server appears to act like a physical server. A virtualized server allows apps to communicate over the Internet. That is, like a physical server, each virtualized server is assigned an Internet address. In one virtualization technology, the match between a virtualized server and a physical server is so accurate that a virtualized server can boot a standard operating system and then allow a user to run standard apps, just as if the operating system runs on the physical server. The point is:

> *To a customer, a virtualized server appears to act like a physical server and apps running on a virtualized server can communicate over the Internet.*

We have seen how using virtualized servers benefits a cloud provider, but if a virtualized server merely acts like a physical server, how does a customer benefit? A few advantages for customers include:

- Ease of creating and deploying new services. Cloud providers and third-party vendors offer software that makes it easy to create new apps for a cloud environment. In addition, they offer software that can deploy an app in the cloud, including software that replicates an app on multiple virtualized servers.

- Rapid scaling of a service. Scaling an app to handle more users means adding more copies. If an app runs on virtualized servers, new copies can be created quickly (e.g., as requests arrive over the Internet).

- Safe and rapid testing of new software or new versions. Before deploying new apps or new versions of software, most large organizations deploy the software on an isolated test system before installing it in production. Virtualized servers allow an organization to create isolated virtualized servers for a test system without interfering with the production systems.

2.7 Business Models For Cloud Providers

As cloud computing emerged, companies were created. Questions arose: what is the best structure for companies engaged in the cloud industry, and how can a company generate revenue? Should a single company handle both physical facilities (air conditioning, power, and building security) and data center facilities, or should separate companies handle the two aspects? Should a single cloud company handle both the cloud hardware (e.g., servers and network equipment) and software (e.g., the software to create and control virtualized servers, the operating systems, and apps running on the servers), or should separate companies handle hardware and software?

A new industry always faces the question of overall structure. For example, before the automobile industry settled on an integrated approach, one set of companies manufactured a chassis and engine and another set of companies built the body and interior of a car. For cloud, it initially seemed some companies would develop the expertise needed to deploy and manage software and others would manage power and air conditioning. To maximize profits, however, companies that initially focused on physical infrastructure began to expand expertise and move into other segments.

To categorize companies in the cloud industry, a classification arose that divides companies into three broad categories. Although they are somewhat loosely-defined and overlap, the categories help clarify the major roles of companies. Each category uses the phrase *as a service*.

- Infrastructure as a Service (IaaS)
- Platform as a Service (PaaS)
- Software as a Service (SaaS)

The next sections explain each of the categories.

2.8 Intrastructure as a Service (IaaS)

In the simplest case, an IaaS company offers physical resources, such as a building, power, and air conditioning. Typically, IaaS companies also provide servers, networking equipment, and basic data storage facilities (e.g., block storage on disk). An IaaS company may offer customers many additional services, such as load balancers, data backup, network security, a way to boot both physical and virtualized servers, and assignment of Internet addresses. An IaaS customer does not need to manage or control the cloud infrastructure. A customer can choose which operating systems and applications run, and may have the ability to control network access (e.g., to configure a firewall). The most advanced IaaS companies use operating systems that can scale the customer's services and the facilities allocated to a customer up or down as needs vary.

2.9 Platform as a Service (PaaS)

The primary goal of PaaS is a facility that allows a customer to build and deploy software in a cloud without spending effort configuring or managing the underlying facility. A company offering PaaS may provide both basic infrastructure and facilities for software development and deployment. Basic infrastructure includes many of the IaaS facilities, such as servers, storage facilities, operating systems, databases, and network connections. Facilities for software development and deployment include compilers, middleware, program libraries, runtime systems (e.g., Java runtime and .NET runtime), and services that host a customer's applications. To emphasize its focus on providing a convenient environment for software development, PaaS is sometimes called *application Platform as a Service (aPaaS)*, and was formerly named *Framework as a Service (FaaS)*, referring to programming frameworks.

Although it is often associated with cloud infrastructure as described above, PaaS can also appear in other forms. For example, some PaaS companies sell software development tools that allow a customer to build and deploy apps on the customer's internal network (i.e., behind the customer's firewall). Other PaaS companies sell a software development tool intended for a cloud environment, but require the customer to obtain servers, storage, network connections and other cloud facilities separately (e.g., from an IaaS company or cloud provider).

2.10 Software as a Service (SaaS)

Software as a Service refers to a subscription model in which a customer pays a monthly fee to use software rather than make a one-time purchase. Cloud computing has enabled the SaaS industry by providing a way for SaaS vendors to scale their offerings to handle arbitrarily many customers.

When a user accesses a SaaS application, the application runs on a server in a cloud data center rather than on the user's computer. Files that the user creates are stored in the cloud data center rather than on the user's local device. Well-known SaaS services include Microsoft's Office 365 in which each customer pays a monthly fee to use programs in the Office suite, such as *Word*, *Excel*, and *PowerPoint*.

SaaS Vendors claim the approach has three advantages:

- Universal access
- Guaranteed synchronization
- High availability

Universal access. SaaS software can be accessed at any time from any device. A user either launches a special app or uses a web browser to access the SaaS site. After entering a login and password, the user can invoke the SaaS app and access a set of files. The universal access guarantee means a user will be able to access the same apps and the same files from any device.

Guaranteed synchronization. The term *synchronization* refers to keeping data identical across multiple devices. With conventional software, synchronization problems arise because a user must load a copy of a file onto a device before using the file. If a user places copies of a file on two devices and then uses one device to change the file, the changes do not automatically appear in the copy on the other device. Instead, a user must manage file synchronization by manually copying the changed version to other devices. We say that the copies can be "out of sync."

The SaaS synchronization guarantee arises because only one copy of each file exists. All changes are applied to a single copy of the file, even if the changes are made using two devices. Consequently, a file created or modified using one device will appear when the user logs in and uses another device; the user will never need to resynchronize the copies across multiple devices.

High availability. Most data centers have uninterruptible power systems that use generators and/or battery backup systems. Thus, the data center can continue operating during a power outage. In addition, the data storage facilities at a data center usually include a backup mechanism, and many store the backups at another physical location. Thus, even if a major catastrophe destroys the data center, a user's data can be recovered from a backup.

Because processing and storage occur in a cloud data center, a device used to access a SaaS service does not need a powerful processor, large memory, or storage. Instead, an access device only needs a user interface and a network connection. Industry uses the term *thin client* to describe such a device†. Some SaaS systems use a web browser as the access app. As a result, SaaS is also called *web-based software*. Other synonyms include *on-demand software* and *hosted software*.

2.11 A Special Case: Desktop as a Service (DaaS)

Many groups have adopted the phrase *as a Service* to describe their particular market segment, including *Network as a Service*, *Security as a Service*, *Disaster Recoverer as a Service*, and *Mobile Backend as a Service* (providing communication between mobile apps and the cloud software they use). One particular form of SaaS stands out as especially relevant to cloud computing. Known as *Desktop as a Service* (*DaaS*), the system implements *remote desktop* access. Like other SaaS offerings, a user runs an app on a local device (i.e., a thin client) that connects the user to DaaS software running in a cloud data center. Instead of providing access to a single app, however, DaaS paints a desktop on the user's screen and allows the user to click on icons, run apps, browse files, and perform other actions exactly as if the desktop was local. The desktop that the user sees, the operating system that supplies the desktop, and the apps a user invokes all run on a server in the cloud instead of the user's local device.

We have already seen the advantages of SaaS systems, and DaaS extends them to all of a user's computing, not just one app. In terms of synchronization, DaaS stores all of a user's files and apps in the cloud data center. So, instead of merely synchronizing data for one particular app, DaaS ensures that all data and all apps remain synchronized. Similarly, all of a user's computing has high availability, and a user can access the desktop from any device at any time.

---

†Of course, a laptop, desktop, or other device can also be used to access a SaaS service, but such devices have more computational power than is necessary.

2.12 Summary

Elasticity forms the key concept behind cloud computing: a customer can lease as many servers as needed, the number can vary over time, and a customer only pays for the number of servers used. Instead of dedicating physical servers to each customer, cloud providers use software technology to create virtualized servers as needed. Multiple virtualized servers can run on a given physical server. Using virtualized servers has advantages for customers as well as for providers.

Companies in the cloud industry can be divided into three broad categories, depending on what services they offer: Infrastructure as a Service (IaaS), Platform as a Service (PaaS), and Software as a Service (SaaS). A special case of SaaS, Desktop as a Service (DaaS), stands out as particularly pertinent to cloud computing. When using DaaS software, a desktop appears on a user's device, but the operating system, apps, and files are all located on a server in the cloud.

Chapter Contents

3

Types Of Clouds And Cloud Providers

3.1 Introduction

The previous chapter explains the concept of elastic computing. The chapter describes how virtualization provides the basic technology that enables a provider to increase or decrease the servers leased by a customer quickly. The chapter also describes the business models used in the cloud industry.

This chapter continues the discussion of the business models. The chapter explains how a large company can use cloud technology internally and how public cloud providers arose to sell cloud services.

3.2 Private And Public Clouds

Cloud computing technologies have been used in two fundamental ways. To describe the two, industry uses the terms:

- Private Cloud – an internal cloud used only by one organization
- Public Cloud – a commercial service used by multiple customers

The next sections examine each of the types along with underlying motivations. Later sections explain how organizations use combinations and variations.

3.3 Private Cloud

We use the term *private cloud* to describe a cloud data center that is owned and operated by an organization and restricted to the organization's computing. Chapter 1 points out that an organization can reduce costs by consolidating all of its servers by placing them in racks in a data center. The organization can further reduce costs by employing cloud technologies. To understand why, suppose three servers have been collected from department A and five from department B. If the servers remain dedicated to departments, a given server may be underutilized or overutilized. More important, if the computing demands of departments vary, one department's servers may be overloaded, while another department's servers are underutilized. When a server becomes oversubscribed, critical business functions can suffer. From an economic point of view, underutilized servers represent unwarranted cost, and the inability to expand facilities quickly represents a risk to business. Of course, an organization must have sufficient resources to handle the peak load, but at other times, the organization can use the elastic approach described in Chapter 2. Instead of physical servers, the approach allows departments to create and assign virtualized servers to balance the load across all physical servers. The point is:

> When an organization uses cloud technology on an internal data center, we say that the organization owns and operates a private cloud. A private cloud can avoid underutilized as well as oversubscribed resources by spreading computing across all physical servers.

3.4 Public Cloud

A company that offers cloud computing to its customers is said to operate a *public cloud*, and the company is known as a *public cloud provider*. For an organization, subscribing to a public cloud constitutes the chief alternative to operating a private cloud internally. When using a public cloud, an organization leases services, including virtualized servers, and then uses the leased facilities to perform computation. We can summarize:

> A company that offers cloud services to customers is known as a public cloud provider, and the cloud facility that such a company operates is known as a public cloud facility.

When using a public cloud, an organization must choose a type of service, as Chapter 2 describes. A medium or large size enterprise customer is likely to choose an IaaS service that offers a set of virtualized servers, an amount of block storage, and Internet connectivity. A large customer may also choose to pay for additional services, such as enhanced data backup, a specialized security service, and database administration. Customers who intend to create and deploy their own apps, may subscribe to a PaaS service that makes it easy to create software and run apps in the provider's cloud. A smaller public cloud customer may choose to subscribe to specific SaaS services, such as a service that runs a web site for the customer or software used to prepare documents.

3.5 The Advantages Of Public Cloud

Why would a corporation choose to use a public cloud provider? Cloud providers advertise three main advantages:

- Economic – much lower cost than a private cloud
- Expertise – access to a staff with expertise on many topics
- Advanced services – offerings not available elsewhere

The economic advantage. Because they serve multiple tenants, cloud providers operate data centers that are much larger than private cloud data centers owned and operated by individual organizations. Therefore, public cloud providers benefit from economy of scale: they can spread costs among more customers, and can negotiate larger discounts on commodity servers and network equipment. For example, instead of conventional (i.e., expensive) network switches, providers use what the industry calls *whitebox switches*. The disadvantage of a whitebox switch is that it only includes basic packet forwarding hardware, and does not provide software that an IT manager can use to configure and control the switch. To operate their whitebox switches, cloud providers employ a technology known as *Software Defined Networking* (*SDN*). A manager connects a computer to each of the switches, and runs SDN control software on the computer. The manager can then interact with the control software to specify the desired configuration, and the control software installs forwarding tables in the whitebox switches that fulfill the manager's specifications.

Discounts on equipment and the use of whitebox network switches means that a public cloud provider spends less on each piece of equipment than a company does when they operate a private data center. In addition, a public cloud provider shares IT staff expertise across all tenants, the cost of staff for a given tenant is less than the cost of staff for a private data center. The savings can be substantial —an employee at Azure claims they can deploy a server for 25% of the cost of a private cloud deployment. The bottom line is:

> *By sharing IT expertise across multiple tenants, negotiating large discounts on commodity servers, and using SDN to control whitebox network switches, a public cloud provider can operate a cloud data center for substantially less than it costs to operate a private cloud data center.*

The expertise advantage. Because it is much larger than the IT department in a single company, a public cloud provider can afford to maintain a staff with specialized expertise. Thus, customers have access to expertise on a much wider range of topics than a local IT staff can provide.

In addition to traditional data center software, such as operating systems, relational database services, and technologies for virtualized servers, cloud providers now include expertise on the broad topic of *Artificial Intelligence* (*AI*) and a technology that AI uses, *machine learning* (*ML*) software.

The advanced services advantage. Initially, providers investigated AI/ML as a way to automate the management of data center hardware and software. For example, by using ML to monitor network traffic, a provider can detect anomalies, such as mis-routed traffic or unexplained spikes in load. Industry uses the term *AIops* to refer to the use of AI/ML to automate data center operations.

More recently, cloud providers have begun offering AIops as a service to customers. In addition to monitoring and automating the deployment and operation of customer apps, providers offer advanced services that use AI/ML software to analyze business data. For example, some AI/ML offerings help companies understand sales trends and can help spot repeat customers.

In addition to AI-based services, providers offer a variety of other advanced services. For example, providers offer services that update a customer's operating systems and other standard software whenever the vendor releases an update. Major cloud providers also offer platforms that allow a customer to build and deploy new apps without understanding the details of the underlying system. In particular, a software engineer can focus on building an app without learning how to create and replicate virtualized servers, establish network connectivity, or handle remote storage. One of the advantages of using such a platform lies in its safety —because the service details deployment details, a customer is less likely to make a mistake that either causes the app to misbehave or incurs excessive cost by consuming data center resources unnecessarily. The point is:

> *Advanced services offered by providers give customers access to AI/ML software and allow customers to create and deploy applications quickly, easily, and safely.*

3.6 Provider Lock-In

Public cloud providers compete to attract new customers and entice existing customers to increase their monthly subscription. One approach focuses on offering specialized services that are not available elsewhere. For example, a provider may offer customers a network management service that allows each customer to monitor and control the network used internally (i.e., between the customer's virtualized servers) and externally (i.e., between a customer's virtualized servers and the rest of the Internet). Because it interacts with the provider's network control system, such a service is inherently linked to the provider.

In addition to advanced services that link to the provider's infrastructure, cloud providers usually offer a *migration service* that makes it easy for a corporate customer to move their computing into the provider's public cloud. To entice customers to switch from another provider, a provider may also offer *cloud-to-cloud migration services*.

To retain customers, providers try to make it difficult for a customer to move to another provider. For example, a provider may offer free or nearly free services that makes it especially easy for a customer to perform routine tasks, such as generating monthly activity summaries, analyzing which of a customer's web pages attract the most attention, or testing new apps in a safe, isolated environment. The catch arises because the service only runs on the provider's infrastructure —if a customer transfers to another provider, the customer will no longer be able to use the service. Even if another provider offers a similar service, the details will change, and a customer must retrain staff for the new system. Industry uses the term *lock-in* to refer to incentives and obstacles that a provider uses to discourage customers from transferring to another provider.

We can summarize:

> *Industry uses the term* lock-in *to refer to the practice of using enticements and obstacles that make it inconvenient or expensive for customers to move to another cloud provider.*

3.7 The Advantages Of Private Cloud

It may seem that the advantages of public cloud —lower cost, access to specialized expertise, and advanced services —will drive all companies to use public cloud services. However, private cloud offers advantages:

- Retention of control and visibility
- Reduced latency with on-premises facilities
- Insurance against future rate hikes

Retention of control and visibility. A public cloud customer does not have access to the underlying infrastructure. If a problem arises, a customer can only report the visible symptoms, but cannot perform root cause analysis because only the cloud provider can examine the network switches and physical servers.

For organizations in a regulated industry, regulations may require the organization to control the placement and transmission of data as well as the hardware and software used. A private cloud makes it possible for such an organization to comply with the rules.

Reduced latency with on-premises facilities. Because they are located within a company, private cloud facilities are said to be *on-premises* (*on-prem*). The delay between employees and a private cloud data center can be significantly lower than the delay to a public cloud, especially if an organization is geographically distant from a public cloud site. An organization with multiple sites may be able to further reduce latency by placing a private cloud data center at each site. If computation and communication remains primarily within a site, the improvement can be noticeable.

Insurance against future rate hikes. Provider lock-in presents a long-term liability for a company. As time goes by, a customer adopts more of the provider's services, and the cost of changing providers increases. A provider can raise rates without fear of losing customers. The point is:

> *Although public cloud seems to offer several compelling advantages, private cloud gives a company more control, can reduce latency, and guards against future rate hikes by a provider.*

3.8 Hybrid Cloud

Instead of choosing between public and private clouds, some organizations opt for a compromise that offers some of the advantages of each. Known as a *hybrid cloud*, the compromise means an organization uses a public cloud provider for some computing, and runs a private cloud for the rest. The balance between the two depends on the organization's needs as well as cost. Examples of why an organization might adopt a hybrid cloud approach include:

- Control when needed
- Computation overflow

Control when needed. We said that some organizations must comply with regulations. For example, a government contractor may need to store classified data on servers with controlled access. Using a hybrid approach allows such a company to enforce restricted access for its private cloud, and reduce overall costs by pushing other data and computations to the public cloud.

Computational overflow. Imagine an organization that uses a private cloud for most of their computing needs. During the organization's peak business season, the private cloud may have insufficient resources to handle the load. Rather than add additional servers, which will remain underutilized most of the time, the organization can send some of the computation to a public cloud during the peak period. To make overflow processing convenient, public cloud providers offer software that helps automate the process. In essence, the software makes the customer's private cloud compatible with the provider's public cloud, meaning that a customer can move overflow from their private cloud to the public cloud with little or no effort. We can summarize:

> *Using a hybrid cloud allows a company to enforce restrictions on data and computations in their private cloud and reduce costs by pushing the rest to a public cloud; the hybrid approach can also be used to handle overflow.*

Of course, the hybrid approach does have some disadvantages. Consider, for example, security management. An organization has complete control over their private cloud and the security systems being used, but will depend on a provider's security systems when using a public cloud. Even if the two systems offer the same functionality, the IT staff must learn to configure and operate two systems.

3.9 Multi-Cloud

Because provider lock-in represents a potential long-term liability, large organizations try to avoid dependence on a single public cloud provider. To take advantage of public cloud services while avoiding lock-in, large organizations adopt an approach that has become known as *multi-cloud*, which means an organization becomes a customer of more than one public cloud provider. The division of computation among providers depends on the structure of the organization and its IT needs. As an example, consider a large corporation with business units. It may be possible to assign some business units to one cloud provider and other business units to another.

Although it avoids lock-in, using multiple cloud providers introduces challenges. It may not be easy, for example, to switch computation or data from one cloud provider to another. Services available from one provider may not match the services available from another. Even if a service is available from multiple providers, the organization may need to write specialized software either to translate from one system to another, or to combine the output from the two providers to generate data for the entire organization.

3.10 Hyperscalers

Industry uses the term *hyperscale* to refer to the largest cloud computing facilities, and the term *hyperscalers* to refer to the companies that own and operate such facilities. Social media and search companies, such as Facebook, Twitter, and Google were among the earliest to build giant data centers to support their respective businesses. Because such companies focus on consumers, early definitions of hyperscale measured the number of simultaneous users data centers could handle. The shift to more general-purpose cloud computing has made the measure of users irrelevant, and the definition has shifted to a measurement of how much a company spends on data center infrastructure. The numbers are huge. By 2018, Google, Amazon, Microsoft, Facebook, and Alibaba were spending about $11B per year just on new servers.

In terms of the public cloud services used by large enterprise organizations, three hyperscalers stand out:

- Amazon's AWS (Amazon Web Services)
- Microsoft's Azure Cloud
- Google' GCP (Google Cloud Platform)

Growth of the cloud industry has been rapid, especially for the top two providers. Initially, Amazon built data centers to support its online retail business. In 2006, before most enterprises were even aware of general cloud computing, Amazon launched AWS. In early 2010, Microsoft released *Windows Azure*, and four years later renamed it *Azure Cloud*. In 2012, cloud services formed an insignificant percentage of revenue for both Amazon and Microsoft. By 2018, just six years later, Microsoft reported that cloud services generated $32.2B or 29% of total revenue. By 2018, AWS generated $25.7B for Amazon or 11% of total revenue.

Hyperscalers are shifting spending to match a shift in services. In 2018, for example, Microsoft revealed it had invested $20B to create its Azure Cloud platform. Now that the data centers have been built, hyperscalers are shifting investment toward AI/ML technologies that will allow them to automate management of their cloud infrastructure and offer advanced services to customers.

3.11 Summary

Cloud computing can be divided into two basic types: public and private. An organization that owns and operates a data center for internal use is said to operate a private cloud, and an organization that owns and operates a data center that external customers use for their computing is said to operate a public cloud.

Because both private and public clouds have advantages, some organizations choose a hybrid approach in which some computing is performed in the organization's private cloud, and other computation is performed in a public cloud. An organization can also adopt a multi-cloud approach in which the organization uses multiple public cloud providers.

Companies that own and operate the largest data centers are known as hyperscalers. Among public hyperscale cloud providers, Amazon's AWS, Microsoft's Azure Cloud, and Google's GCP have attracted enterprise customers.

Part II

Cloud Infrastructure And Virtualization

The Underlying Mechanisms And The Virtualizations That Make Them Useful

Chapter Contents

4

Data Center Infrastructure And Equipment

4.1 Introduction

The first part of the text presents the general concept of cloud computing and its use in private and public clouds. This chapter begins a deeper look by describing data center infrastructure and the equipment used, including cooling systems, which have become a critical part of data center design. Remaining chapters in the section discuss virtualization and its role in cloud computing.

4.2 Racks, Aisles, And Pods

Physically, a data center consists of a building, or part of a building, that houses the equipment. Often, a data center occupies a single large open area without walls. Like a giant retail store, columns are spread throughout the area to support the ceiling.

A data center can be huge. Flexential's ComPark data center in Denver comprises 148,000 square feet of space in a single large room. The distance from one end to the other is 710 feet, more than two football fields laid end to end. The largest data centers, especially the ones operated by media companies, can exceed 1,000,000 square feet.

Physically, racks holding equipment are placed side by side in rows, leaving *aisles* between them. Logically, however, a data center is not merely composed of long rows of racks. Instead, a data center is built by replicating a basic set of equipment known as a *pod*, sometimes written *PoD* for *Point of Delivery*†. In addition to servers, a pod in-

---

†At least one vendor uses the acronym *POD* for a product, adding confusion to the interpretation.

cludes storage and networking facilities plus *Power Distribution Units* (*PDUs*) that deliver electrical power to the pod; a pod may include management facilities used to configure, monitor, and operate the pod.

Some vendors sell pre-built pods in which equipment has been mounted in racks, the racks have been joined together, and the unit is ready to be shipped to a data center and moved into place. Once the owner connects power, air conditioning, and network connections, the pod is ready to function. To summarize:

> *Rather than installing individual racks and mounting individual servers, a data center owner can acquire and install a pre-built pod.*

4.3 Pod Size

How large is a pod? Data center owners have experimented with a variety of sizes. An early design created pods with over 200 racks per pod. The industry has moved to smaller sizes, where a pod with 48 racks is considered "large," and an average-size pod contains 12 to 16 racks.

Although shipping considerations favor smaller pre-built pods, three other factors have motivated the change:

- Incremental growth
- Manageability
- Power and cooling

Incremental growth. Choosing a smaller pod size allows a data center owner to grow the data center continuously in small increments rather than waiting until a large pod is justified.

Manageability. Management tools allow a data center owner to manage each pod independently. A smaller pod size makes it easier to find and repair problems, and keeps problems contained within a pod.

Power and cooling. Interestingly, electrical power and cooling have become a major consideration when choosing a pod size. The next section explains why.

4.4 Power And Cooling For A Pod

Power consumption and cooling dominate many aspects of data center design because data centers consume huge amounts of power. For example, the Inner Mongolia Information Park owned by China Telecom consumes over 150 Megawatts of electric power. To put the number in perspective, the Yale Environment 360 notes that 150 Megawatts is approximately the amount of electrical power consumed by a city of a million people. Together, data centers consume over 2 percent of the world's electricity, and by some estimates, data center power usage will grow to almost 25 percent of the world's electricity by 2025.

How are pod size and power consumption related? Each pod has a dedicated electrical feed, which means the size of a pod determines the amount of power the pod requires. Industry uses the informal term *power density* to refer to the power needed for a rack, and observers have noted that power density has increased over the past years as servers gained more cores and network equipment moved to higher data rates. Therefore, the power needed by a pod has increased. One way to limit the power consumed by a given pod consists of reducing the pod size.

Cooling is the counterpart to power. As it consumes power, electronic equipment produces heat. To prevent it from malfunctioning and eventually burning out, the equipment must be cooled. Consequently, as power consumption increased, the need for cooling also increased. When we think of data center equipment, it is easy to focus on servers and network switches. However, heat removal has become so significant that the equipment used to cool electronics has become as important as the electronics themselves.

To summarize:

> *Because the electronic systems in a data center generate substantial amounts of heat, the systems that remove heat have become major parts of the data center infrastructure.*

Designers have invented several ways to reduce heat, including:

- Raised floor pathways and air cooling
- Thermal containment and hot/cold aisles
- Exhaust ducts (chimneys)
- Lights-out data centers

The next sections describe each of the techniques.

4.5 Raised Floor Pathways And Air Cooling

Data centers employ a design pioneered by IBM in the 1950s for mainframe computers: *raised floor*. A metal support structure supports a floor from one to four feet above the concrete floor of the building. The space between the real floor and the raised floor can be used to hold power cables and for air cooling. Large outdoor air conditioning units (compressors) surround the building, and pipes carry compressed refrigerant to indoor units where it is used to chill air. The chilled air is then forced under the raised floor.

A raised floor consists of square tiles that are 60 cm (approximately 17 inches) on a side. Most tiles are solid, but the tiles under racks are perforated, allowing air to flow upward into the racks. Racks are, in turn, designed to allow the chilled air to flow up the sides of the rack. Each piece of equipment in a rack contains a fan that blows the cool air through the unit, keeping the electronics cool. That is:

> *Current data center designs push chilled air under a raised floor. Perforated floor tiles under each rack allow the chilled air to flow upward through the rack to cool equipment.*

4.6 Thermal Containment And Hot/Cold Aisles

Hot air leaves each piece of equipment, venting into the data center. Overall air flow in the data center must be designed carefully to move hot air away from the racks, ensuring that it cannot be accidentally drawn back into another piece of electronic equipment. In particular, a designer must ensure that the hot air leaving one piece of electronic equipment is not pulled into another piece.

A technique known as *thermal containment* (or *aisle containment*) offers one solution. The idea is to direct heat to contained areas. To use the approach, an owner places a cover on the front of each rack and leaves the back of the rack open. Equipment is arranged to pull cold air from the sides of the rack and vent hot air out the back. Containment can be further enhanced by orienting racks to create an aisle with the fronts of racks facing each other and then an aisle with the backs of racks facing each other. Because racks exhaust hot air out the back, the aisle with backs of racks facing each other constitutes a "hot" aisle, and the next aisle with the fronts of racks facing each other constitutes a "cold " aisle. Figure 4.1 illustrates the concept by showing rows of racks arranged to create alternating hot/cold aisles. Gray arrows indicate the flow of hot air from the back of each rack.

To keep hot air from collecting near the racks, a set of fans in the ceiling pulls the hot air upward and sends it through ducts back to the indoor air conditioning units. Thus, air circulates throughout the data center continuously. Cool air is forced under the floor and upward through the racks, absorbing heat from the equipment. The heated air is drawn upward and sent back to the indoor air conditioning units, where the heat is removed and the cool air is pumped under the raised floor again.

4.7 Exhaust Ducts (Chimneys)

Despite fans in the ceiling that draw hot air upward, the temperature near racks with high power density can be higher than other areas of a data center. Designers refer to such areas as *hot spots*. Designers employ a variety of techniques to reduce the temperature near hot spots, including leaving some slots in each rack empty, and leaving a completely empty rack between two especially hot racks. For areas that generate inordinate heat, a vertical duct with a fan can be placed over the area with a fan to move hot air upward. Industry uses the informal term *chimney* to refer to such ducts.

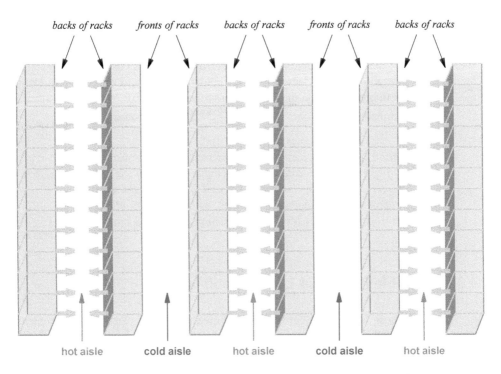

Figure 4.1 An illustration of hot/cold aisles that aid in thermal containment by keeping heated air from one piece of equipment from flowing into another piece of equipment.

Placing a vertical duct over a hot spot is especially pertinent to pods because the center of a pod tends to form a hot spot. Recall that some vendors offer pre-built pods. A pre-built pod may include a vertical duct to move hot air upward, away from the pod.

4.8 Lights-Out Data Centers

An operational paradigm has been invented that helps reduce heat in a data center: minimize any extraneous use of electricity. In particular, avoid keeping an area lit when no humans are working in the area. Known as a *lights-out data center*, the scheme means that entire parts of the data center operate in the dark. To minimize the time lights must be on, servers, network switches, and storage equipment are accessed and managed over a network.

The availability of reliable, automated failure recovery and maintenance systems has further enabled the lights-out approach. Automated systems are used for routing, monitoring tasks, and handling fast cut-over during failures. For example, we will see that each network switch in a data center has at least two connections that can be used to reach other switches or the Internet. If one connection fails, automated software detects the failure and changes forwarding so network packets flow over the other connection. As a result, humans only need to enter the data center to replace physical equipment and handle unusual situations.

In addition to reducing energy costs, The lights-out approach has three advantages. Using automated systems to monitor a data center offers owners cost savings by reducing the staff size; automation is less likely than human operators to misconfigure equipment; and restricting personnel in the data center reduces the threat of malicious attacks.

We can summarize:

> *In addition to thermal containment and vertical ducts over hot spots for cooling, some data centers follow a* lights-out *approach in which a data center operates in the dark. As well as reducing energy costs, using automated management systems with a lights-out approach reduces staff costs and avoids both human error and malicious attacks.*

4.9 A Possible Future Of Liquid Cooling

Although most data centers currently use chilled air to cool electronic components, power density continues to increase. For example, a pod that currently consumes 600 Kilowatts of electrical power, may soon need more cooling. Unfortunately, merely pumping chilled air through electronic equipment can only remove a limited amount of heat in a given time. Decades ago, supercomputer designers found a better way to remove heat from electronic circuits: *liquid cooling*. The physics behind liquid cooling is straightforward: a chilled liquid passing over electronic components will absorb more heat than chilled air.

Supercomputers that use liquid cooling contain hydraulic systems that pump cool liquid across each of the electronic circuits. In fact, the equipment in a liquid cooling system operates like the equipment in an air cooling system with an outdoor unit to compress refrigerant that is pumped through the equipment. Of course, submerging electronic components in liquid will cause problems if the liquid conducts electricity. So, the refrigerant used with liquid cooling is non-conductive, allowing it to touch electrical parts without a problem.

When it changes from air cooling to liquid cooling, a data center must install hydraulic equipment to circulate cold liquid refrigerant to the racks and return heated refrigerant to the cooling unit. In addition, all servers and network equipment must be replaced with units that have hydraulic fittings to accommodate liquid cooling. The point is:

> *Although it can remove more heat than air cooling, changing to liquid cooling requires an owner to replace most of the data center infrastructure.*

4.10 Network Equipment And Multi-Port Server Interfaces

Network connectivity forms the second most important service that cloud providers offer. A *network switch* in each rack connects to each server in the rack and provides communication among the servers as well as communication to the rest of the data center and the Internet. Data center switches use *Ethernet* technology, and the switches are sometimes called *Ethernet switches*.

In most data centers, network cables run along open cages suspended above the racks. The switch in each rack is usually placed near the top, giving rise to the term *Top-of-Rack switch* (*ToR switch*)†. To permit rapid data transfers, the connections between the ToR switch and each server must operate at high speed. Initially, data centers used 1 *Gigabit per second* (*Gbps*) Ethernet leading to the name *GigE*. More recently, data centers have used 10 Gbps and 40 Gbps technologies.

To further increase the rate at which data can be sent, each server can use a *multi-port network interface card* (*multi-port NIC*). Each of the ports connects to the ToR switch, and each operates independently and in parallel. Thus, a NIC with K ports means the server has K network interfaces. allowing it to send and receive K times as much data each second as a single network interface. A multi-port NIC works well with a multi-core server because it allows the server to send and receive more data.

4.11 Smart Network Interfaces And Offload

Network processing requires computation. Each time a packet arrives at a server, fields in the packet headers must be examined to determine whether the packet has been formed correctly and whether the packet is destined to the server. Similarly, each time a packet is sent, headers must be added to the data. In a conventional computer, software in the operating system performs all packet processing tasks, which means the processor spends time handling each packet.

The processing required for network packets can be significant, especially if the data must be encrypted before being sent and decrypted when received. To send and receive data at high speed, the network interfaces used in many data centers contain special hardware that handles packet processing tasks. Known informally as a *smart NIC*, such an interface card can assemble an outgoing packet, compute a checksum, and even encrypt the data. Similarly, a smart NIC can check the headers on each incoming packet, validate the checksum, extract and decrypt the data, and deliver the data to the operating system. We say that a smart NIC *offloads* processing, which has two advantages:

> *Using a smart NIC to offload packet processing means the network can operate at full speed and instead of spending time on packet processing tasks, a processor can be devoted to users' computations.*

---

†Other equipment, such as electrical power converters, may also be placed near the top of a rack to reduce the heat generated near electronic equipment.

4.12 North-South And East-West Network Traffic

Recall that each rack in a data center contains a Top-of-Rack (*ToR*) network switch that connects to each of the servers in the rack. Two questions arise: how should the ToR switches in all the racks be interconnected to form a network in the data center, and how should the data center network connect to the Internet?

A variety of network architectures have been used in data centers. Although many variations exist, we can group them into two broad categories based on the type of traffic they are intended to handle:

- North-south traffic
- East-west traffic

North-south traffic. Industry uses the term *north-south traffic* to describe traffic sent between arbitrary computers on the Internet and servers in a data center. Recall, for example, that early data centers focused on large-scale web sites. Web traffic falls into the category of north-south traffic.

How should a data center network be designed to handle web traffic? In theory, the data center could use the arrangement that Figure 1.1† depicts —a single load balancer dividing incoming requests among all servers in the data center. In practice, however, a network device that performs load balancing has a limited number of connections. Therefore, to accommodate large scale, a network must be designed as a hierarchy with an initial load balancer dividing requests among a second level of load balancers, as Figure 4.2 illustrates.

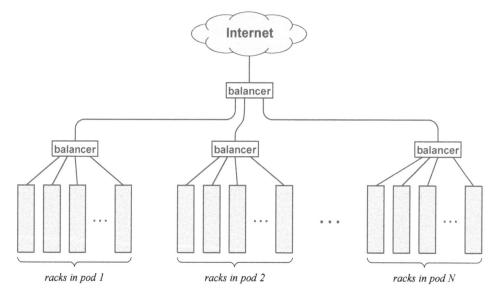

Figure 4.2 Illustration of a simplified hierarchy of load balancers used to achieve a large-scale web service.

---

†Figure 1.1 can be found on page 9.

An actual network is much more complex than the simplistic version in the figure. For example, instead of using specialized load balancer devices, the load balancing function is built into network switches. Furthermore, to ensure that the failure of one connection does not leave some servers cut off, data center networks are designed with redundant connections among switches. Nevertheless, the figure helps explain the terminology. In a data center, network connections run above the racks. Therefore, the figure has been drawn with the Internet on top and racks of servers on the bottom, meaning that network traffic flows vertically. If one thinks of the figure as a map, requests arriving from the Internet flow from north to south, and replies flow from south to north. Hence, web traffic is classified as *north-south* traffic.

East-west traffic. As data centers moved from large-scale web service to cloud computing, network traffic patterns changed. Consider a company using cloud computing. When the company fills an order, software may need to access both a catalog of products as well as a customer database. Similarly, when a manager approves time off, software may need to access an employee's record, payroll data, and the company's accounting system. Communication within the company means network traffic will travel among the servers the company has leased (i.e., from a server in one rack to a server in another). In terms of Figure 4.2, communication proceeds left and right, which leads to the name east-west traffic.

4.13 Network Hierarchies, Capacity, And Fat Tree Designs

Arranging a data center network as a hierarchy has a disadvantage: links near the top of the hierarchy carry more traffic than links farther away. To understand capacity, look again at Figure 4.2. The link between the Internet and the first load balancer carries 100% of the traffic that enters and leaves the data center. At the next level down, however, the traffic is divided among the pods. If the data center contains P pods, each of the P links that connects the first load balancer and the load balancer for a pod only needs to carry $1/P$ of the data. If a pod contains R racks, a link between a rack and the load balancer for the pod only needs to carry $1/RP$ of the data, as Figure 4.3 illustrates.

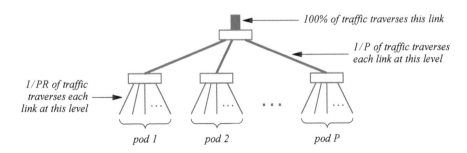

Figure 4.3 Internet traffic in a conceptual network hierarchy for a data center with P pods and R racks per pod.

The networking industry uses the colloquial term *pipe* to refer to a network connection, and talks about data *flowing* through the pipe, analogous to the way a liquid flows through a conventional pipe. The term fat pipe refers to a network connection that has high capacity. The terminology has been extended, and a hierarchy similar to the one in Figure 4.3 is known as a *fat tree†*. In fact, the terminology is somewhat misleading because links have successively less capacity as one moves down the hierarchy.

4.14 High Capacity And Link Aggregation

For a data center that handles high volumes of traffic, links near the top of the hierarchy can require extremely high capacity. A data center owner must face two constraints:

- Capacities available commercially
- Cost of high-capacity network hardware

Capacities available commercially. Network hardware is not available in arbitrary capacities. Only specific capacities have been standardized. Ethernet hardware is available for 1, 10, 40, 100, and 400 Gigabits per second (*Gbps*), but not 6 Gbps. Thus, a hierarchy must be designed carefully to use combinations of capacities that match commercially available hardware.

Cost of high-capacity network hardware. A second factor that must be considered is the cost of network hardware. The cost of high-capacity network hardware is significantly higher than the cost of hardware with lower capacity, and the cost disparity is especially high for networks that cover long distances.

The above factors have led to a technology, known as *link aggregation* or *bonding*, that sends data over multiple low-capacity networks simultaneously. Figure 4.4 illustrates the idea by showing how ten inexpensive 10 Gbps networks can be used to achieve a network with a capacity of 100 Gbps.

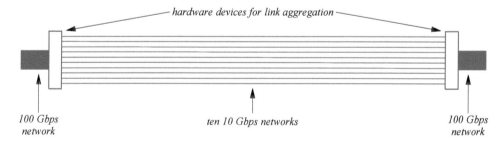

Figure 4.4 Illustration of two hardware devices aggregating ten networks operating at 10 Gbps to provide a 100 Gbps network.

---

†A hierarchy is a *tree* in the graph-theoretic sense because it does not contain a cycle.

Although it can be used to create higher capacity network connections, bonding does have some drawbacks. Using multiple lower-capacity networks introduces more possible failure points. Furthermore, hardware is only available for certain combinations of network capacities. Thus, industry has moved away from the traditional hierarchical approach that Figure 4.2† depicts.

4.15 A Leaf-Spine Network Design For East-West Traffic

How can a data center network be designed that handles large volumes of east-west traffic without using a hierarchical design? The answer lies in parallelism and a form of load balancing. The specific approach used in data centers is known as a *leaf-spine network architecture‡*. In leaf-spine terminology, each Top-of-Rack switch is called a *leaf*. The data center owner adds an additional set of *spine* switches and connects each leaf switch to each spine switch. Figure 4.5 illustrates the arrangement with an example that has six leaf switches and four spine switches.

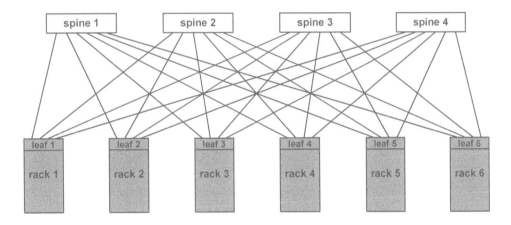

Figure 4.5 An example leaf-spine network with a leaf (i.e., a ToR switch) in each rack connecting to four spine switches.

The leaf-spine architecture offers two main advantages over a hierarchical design:

- Higher capacity for east-west traffic
- Redundant paths to handle failures

Higher capacity for east-west traffic. To understand the capacity, consider traffic traveling east-west from one rack to another. Because both the source and destination racks connect to all four spine switches, four independent paths exist between each pair

---

†Figure 4.2 can be found on page 44.

‡The architecture is also known as a *folded Clos network* because the idea was invented by Charles Clos in 1938 for use in the telephone system.

of racks, one path through each spine switch. Furthermore, a leaf switch equipped with *Equal Cost Multipath Routing* (*ECMP*) technology can be configured to divide traffic equally among the paths.

Consider how ECMP can be used in the network in Figure 4.5. As an example, suppose servers in rack 1 are sending data to servers in rack 6. ECMP means one-fourth of the data will travel through spine 1, another fourth of the data will travel through spine 2, and so on. The ability to divide data among paths leads to an important property of the leaf-spine architecture: capacity can be increased incrementally.

A leaf-spine architecture allows a data center owner to increase network capacity incrementally merely by adding more spine switches.

Redundant paths to handle failures. To understand how leaf-spine accommodates failure, consider Figure 4.6 which illustrates the leaf-spine configuration in Figure 4.5 after spine switch 3 has failed.

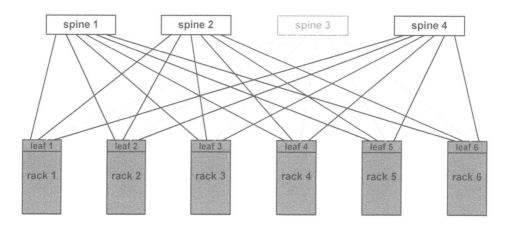

Figure 4.6 An example leaf-spine network in which the third spine switch has failed.

As the figure shows, removing a single spine switch reduces capacity, but does not disrupt communication because three paths still remain between any pair of racks. Of course, packet forwarding in the switches must be changed to accommodate the failure by dividing traffic among the spines that remain functional. Once the hardware in a leaf switch detects that a spine is no longer available, network management software performs the reconfiguration automatically, without requiring a human to handle the problem. Similarly, when the spine switch has been replaced and links become active, network management software will automatically reconfigure forwarding to use the spine again.

4.16 Scaling A Leaf-Spine Architecture With A Super Spine

Although it works for small numbers of racks, connecting all racks to each spine switch does not scale to tens of thousands of racks because the largest switches do not have tens of thousands of ports. To handle scaling, a data center uses a separate leaf-spine network to connect the racks in each pod. An additional level of switches known as a *super spine* is added to connect the spine switches to each pod. Each super spine switch connects to every spine switch. Figure 4.7 illustrates the arrangement.

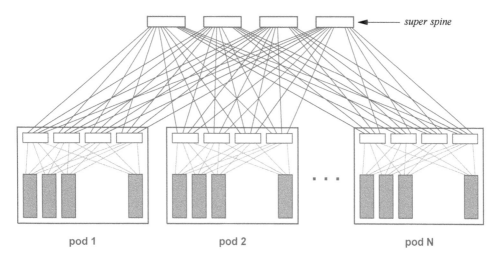

Figure 4.7 Illustration of a super spine configuration in which each pod has a leaf-spine network, and each spine switch connects to each of the super spine switches.

4.17 External Internet Connections

How does a super spine architecture connect to the Internet? The incoming Internet connection may pass through a router, a hardware firewall, or other equipment, and eventually reaches a special switch. In fact, most data centers dedicate at least two switches to each connection in case one switch fails. The two external connection switches each connect to all the super spine switches as if they are spine switches (i.e., the two act like a miniature pod).

The advantage of the super spine architecture should be clear: short paths across the data center for both internal and external traffic without requiring high-capacity links. When two servers in a rack communicate, packets flow from one to the other through the Top-of-Rack (i.e., leaf) switch. When two servers in racks in the same pod communicate, packets flow through the sender's leaf switch to a spine switch, and through the receiver's leaf switch to the receiver. Finally, when a server communicates with a distant server or an external site, traffic flows across the super spine. For example, when a server in one pod communicates with a server in another pod, packets flow from the sender through the sender's leaf switch to a spine switch in the sender's pod, to a super spine, to a spine switch in the receiver's pod, and through the receiver's leaf switch to the receiver.

4.18 Storage In A Data Center

Data center providers follow the same basic approach for storage facilities as they do for computational facilities: parallelism. That is, instead of a single, large disk storage mechanism, the physical storage facilities used in data centers consist of many inexpensive, commodity disks. Early data centers used electromechanical disks, sometimes called *spinning disks*. Modern data centers use *Solid State Disks* (*SSDs*).

Although the shift to solid state disk technology has increased reliability, the most significant change in data center storage arises from a change in the placement of storage equipment. To understand the change, recall that many early data centers were designed to support large-scale web service. Data centers used conventional PCs that each had their own disk. Data for the web site was replicated, and each server had a copy on the local disk. Only transactions, such as placing an order, required access to a central database.

The introduction of multi-tenant cloud systems made local storage on servers problematic for two reasons. First, a given server ran virtualized servers from multiple customers at the same time, and software was needed to limit the amount of storage each customer could use. Second, because disks were spread across the entire data center, replacing a failed disk required sending a staff member to the correct rack†.

To overcome the problems, data centers employ an approach for storage analogous to the approach used for computation: virtualized disks. In a small data center, the owner places all physical storage devices in a centralized location; in a larger data center, multiple locations are used. Software then creates virtualized disks for customers. When a virtualized server is created, the virtualized server is granted access to a corresponding virtualized disk. As software on a virtualized server accesses or stores data on its disk, requests travel across the data center network to the storage facility, and replies travel back over the network. Industry uses the term *block storage* to refer to virtualized disks because traditional disks provide an interface that transfers a fixed-size *block* of data.

From the point of view of software running on a server, *block storage* behaves exactly like a local disk except that transfers must pass across a network. Network communication introduces latency, which means that accessing block storage over a network can take longer than accessing a local disk. To avoid long delays, a data center owner can create multiple storage facilities and place each facility near a group of racks. For example, some data centers place a storage facility with each pod. The higher reliability of solid state disks has lowered failure rates, making replacement much less frequent.

---

†The spinning disks used in early cloud data centers failed more often than solid state disks.

4.19 Unified Data Center Networks

Early data center storage facilities used specialized network hardware that was designed to optimize remote storage access. The specialized hardware was expensive and required running extra cables. Interestingly, the use of specialized hardware is fading, and data centers are now using a single network for storage access as well as other communication. Ethernet hardware emerged that offers both high capacity and relatively low cost, and the leaf-spine architecture emerged that supports east-west traffic (storage access is a special form of east-west traffic). Consequently, data centers have moved to a unified network that handles all traffic.

The point is:

> *The availability of low-cost Ethernet hardware and a leaf-spine network architecture has allowed data centers to eliminate special-purpose storage networks and move to a single, unified network that carries storage access traffic as well as other traffic.*

4.20 Summary

A data center consists of racks arranged in aisles and divided logically into pods. Although we think of data center facilities as servers, networks, and storage hardware, physical infrastructure includes power, power backup, and air conditioning facilities. Raised floor allows cool air to reach each rack, and techniques such as hot/cold aisles and lights-out data centers help manage cooling and power consumption.

Data center servers employ multi-port and smart network interface cards that offer high speed and offload network processing. Early data centers used hierarchical networks in a fat tree design that requires highest capacity near the top of the tree. Modern cloud centers use a leaf-spine network architecture that supports east-west traffic (i.e., traffic among racks in the data center).

Early data centers used conventional computers that each had a local disk. A modern data center places all physical storage hardware in a centralized location and provides access to virtualized block storage over a network. To minimize the latency between a server and block storage, a large data center can choose to place storage facilities at multiple locations (e.g., one per pod). Data centers originally had specialized networks used to connect servers to storage facilities. The emergence of low-cost Ethernet hardware that offers high capacity and the use of leaf-spine network designs has motivated a change to a single, unified network that carries storage access traffic as well as other traffic.

Chapter Contents

5

Virtual Machines

5.1 Introduction

The previous chapter describes the basic infrastructure of a data center, including physical infrastructure, such as power and cooling, as well as racks, pods, servers, networks, and storage facilities. This chapter begins the description of virtualization mechanisms and technologies that allow a cloud owner to use the facilities to provide elastic computing. The chapter explores the concept of a Virtual Machine and the support systems needed to create and manage Virtual Machines. The next chapters describe an alternative virtualization technology used in data centers, containers. Subsequent chapters consider virtual networks and virtual storage facilities.

5.2 Approaches To Virtualization

The concept of virtual machines existed long before cloud computing was invented. The technologies used to implement virtual machines can be divided into three broad categories:

- Software emulation
- Para-virtualization
- Full virtualization

Software emulation. The idea of using software to emulate a computer has many applications. One use involves accommodating heterogeneous computers. Suppose a user who has a particular type of computer, C_1, wants to run a program, P, that has been compiled for another type of computer, C_2. A piece of software called a *software emulator* solves the problem. The emulator treats program P as data by repeatedly reading an instruction, doing whatever C_2 would do, and then moving to the next instruction.

The software emulation approach has been especially popular as a way to make a new programming language available quickly. Before a language can be used on a particular type of computer, someone must write a compiler that translates programs into instructions the computer understands. Writing a compiler is time-consuming and error prone, and often limits how quickly a new language can be widely adopted. Software emulation allows a language designer to write one compiler and then use the compiler on all types of computers.

To implement a compiler for a new programming language, use an existing language to build two tools: a compiler and a software emulator. To simplify writing the tools, define a hypothetical computer (i.e., a type of virtual machine). Build a compiler that translates programs written in the new language into code for the hypothetical machine. Build a software emulator that reads code for the hypothetical machine and carries out the steps the code specifies. The community uses the term *byte code* to describe instructions for the hypothetical computer, calls the compiler a *byte code compiler*, and calls the emulator a *byte code interpreter*. Once the compiler and software emulator have been written, porting them to a new type of computer is much easier than building a new compiler†.

Although it offers the advantages of being general and making it easy to move apps to new types of computers, software emulation has a major disadvantage that makes it unattractive as the primary form of virtualization used in data centers: software emulation adds significant overhead resulting in slow execution. The point is:

> *Although it increases the portability of compilers and other apps, software emulation incurs significant execution overhead, making it ineligible as a primary form of virtualization in a data center.*

Para-virtualization. Now classified as *para-virtualization*, an approach to virtualization pioneered in the 1960s is designed to allow multiple operating systems to run on a single computer at the same time. Of course, the virtualization mechanism cannot allow a given operating system to take over the underlying hardware completely. Instead, para-virtualization uses a piece of software known as a *hypervisor* to control multiple operating systems and move the processors among them analogous to the way a single operating system moves the processor among a set of running processes.

Unlike software emulation, para-virtualization allows software to run at high speed. That is, para-virtualization avoids the inherent overhead of software emulation and allows the processor to execute instructions directly with no extra software involved. We say that para-virtualization allows instructions to execute *natively*. A later section describes how native execution is possible.

To prohibit a given operating system from commandeering the hardware, para-virtualization requires the code for an operating system to be altered before it can be

---

†The Java Programming Language uses the software emulation approach, and defines a *Java Virtual Machine (JVM)*.

used. All *privileged instructions* (i..e., instructions used to gain control of the hardware and I/O devices) must be replaced by calls to hypervisor functions. Thus, when an operating system tries to perform I/O, control passes to the hypervisor, which can choose whether to perform the requested operation on behalf of the operating system or deny the request.

> *An early form of virtualization known as* para-virtualization *allows multiple operating systems to run on a computer at the same time by using a piece of software known as a* hypervisor *to control the operating systems. Para-virtualization has the advantage of allowing high-speed execution and the disadvantage of requiring code to be altered to replace privileged instructions before it can be run.*

Full virtualization. Like para-virtualization, the approach known as *full virtualization* allows multiple operating systems to run on the same computer at the same time. Furthermore, full virtualization avoids the overhead of software emulation and allows operating system code to run unaltered. The next sections explain full virtualization in more detail. For now, it is sufficient to know:

> *Full virtualization allows multiple operating systems to run on a single computer at the same time and at high speed without requiring operating system code to be altered.*

5.3 Properties Of Full Virtualization

The full virtualization technologies currently used to support Virtual Machines (VMs) in cloud data centers have three key properties:

- Emulation of commercial instruction sets
- Isolated facilities and operation
- Efficient, low-overhead execution

Emulation of commercial instruction sets. To a customer, a VM appears to be identical to a conventional computer, including the complete instruction set. Code that has been compiled to run on a commercial computer will run on a VM unchanged. In fact, a VM can boot and run a commercial operating system, such as *Microsoft Windows* or *Linux*.

Isolated facilities and operation. Although multiple VMs on a given server can each run an operating system at the same time, the underlying system completely isolates each VM from the others. From the point of view of an operating system running on a VM, the operating system thinks it controls all of physical memory, a set of I/O devices, and the processor, including all cores.

Efficient, low-overhead execution. The most important and surprising aspect of VM technology arises from the low overhead a VM imposes. In fact, until one understands how a VM works, performance seems impossibly fast because most instructions, especially those used in application code execute natively (i.e., at full hardware speed). To summarize:

> *When an application runs on a VM, most instructions execute as fast as they do when the application executes directly on the underlying processor.*

The next sections explain the conceptual pieces of software that a VM system uses and how the implementation can achieve such high performance.

5.4 Conceptual Organization Of VM Systems

The general idea behind VM is straightforward: load software onto a server that allows the cloud provider to create one or more VMs. Allow the tenant who owns each VM to boot an operating system on the VM, and then use the operating system to launch one or more applications.

The key piece of software responsible for creating and managing VMs is known as a *hypervisor*. We think of a hypervisor as controlling the underlying hardware. Each VM the hypervisor creates is independent of other VMs. Figure 5.1 illustrates the conceptual organization of a server running hypervisor software, and a set of VMs that each run an operating system and apps.

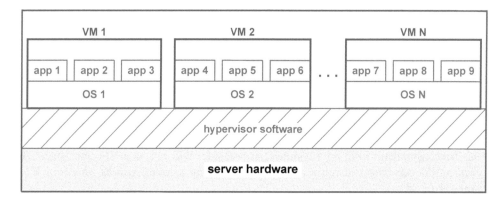

Figure 5.1 Illustration of a server running a hypervisor† and a set of VMs.

---

†Technically, the figure shows a *type 1 hypervisor*, the type commonly used in data centers; a later section describes a *type 2 hypervisor* that an individual might run on their computer.

5.5 Efficient Execution And Processor Privilege Levels

The question arises, how can an application running in a VM operate at hardware speed? Figure 5.1 makes it appear that two layers of software separate a running application from the underlying hardware: an operating system and a hypervisor. We know that software operates much slower than hardware, so how is high speed possible?

To understand how a VM runs apps at hardware speed, consider how an operating system runs apps on a conventional computer. When a user launches an app, the operating system loads the code for the app into the computer's memory. The operating system then instructs the processor to start executing the code. Execution proceeds at the hardware rate because the processor executes code for the app directly without going "through" the operating system.

An application cannot be allowed to execute all possible instructions or the computer would be vulnerable to hackers who might steal information or use the computer in a crime. To prevent such problems, the processor hardware used in a conventional computer has two *privilege levels* or *modes of operation*. Operating system code runs in *kernel mode*, which allows the operating system to perform all possible instructions. When it switches the processor to application code, the operating system also changes to *user mode*, which means only basic instructions are available. If the application makes a *system call* to request an operating system service (e.g., to read from a file), the processor transitions back to kernel mode. Figure 5.2 illustrates operating system and app code in memory and some of the possible transitions.

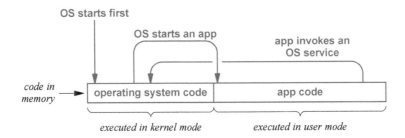

Figure 5.2 Illustration of operating system and app code in memory. The processor executes each at the same high speed, changing mode when transitioning from one to the other.

User mode means an app can access memory that has been allocated to the app, and the app can perform basic instructions, such as addition or subtraction. However, an app cannot access memory owned by the operating system or the memory owned by other apps. If an app attempts to make an instruction that requires privilege or attempts an illegal memory access, the processor raises an exception, which transfers back to a special point in the operating system to allow the operating system to handle the problem.

5.6 Extending Privilege To A Hypervisor

When a server runs a hypervisor and VMs, the software uses the same approach as an operating system uses to run apps, except that the processor employs three levels of privilege: one for the hypervisor, a second for an operating system, and a third for apps. Only the hypervisor can create a VM and allocate memory to the VM. The operating system is restricted to the memory that has been allocated to its VM. The operating system can run apps in the memory it has been allocated. As with a conventional computer, the processor always runs code directly from memory, allowing the code to be executed at hardware speed.

Figure 5.3 illustrates software in memory when a VM runs and shows some of the transitions among privilege levels as a hypervisor starts a VM, the operating system in the VM launches an app, the app calls an operating system service, and the operating system eventually exits (i.e., shuts down the VM).

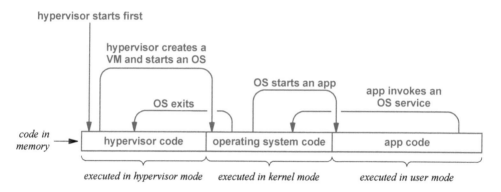

Figure 5.3 Illustration of transitions among code in memory when a hypervisor runs a VM with an operating system and the operating system runs an app.

5.7 Levels Of Trust

We can think of the three processor modes as imposing three levels of trust. When it runs in hypervisor mode, the processor can perform any operation available on the hardware. Thus, the hypervisor code is trusted completely. When it runs in kernel mode, the processor restricts the set of operations to ensure that the operating system cannot affect other VMs or the hypervisor. Because the hardware mode prevents interferences, the operating system code does not need to be trusted as much as hypervisor code. Finally, when it runs an application in user mode, the processor restricts the set of operations even further, making it impossible for an app to affect other apps or the operating system. Because the hardware prevents interference, an app does not need to be trusted as much as an operating system.

Of course, each entity in the trust hierarchy need to be trusted to manage entities under it. When it creates a VM, a hypervisor must be trusted to configure the hardware correctly to ensure the VM is isolated from other VMs. Similarly, when an operating system runs an app, the operating system must be trusted to configure the hardware to keep the app protected from other apps. If an entity attempts to perform an operation that exceeds its trust level, control passes up to the entity at the next trust level. We can summarize:

> *The software used for VMs runs a trust hierarchy controlled by processor modes. A hypervisor is trusted to establish and manage VMs, an operating system is trusted to create and manage a set of apps, and an app runs with the least trust. An attempt to exceed the trust boundary causes the processor to return control to the software at the next higher trust level.*

5.8 Levels Of Trust And I/O Devices

Adding an extra level of privilege makes I/O devices problematic. To understand why, consider how an operating system manages I/O devices (e.g., a screen, keyboard, disk, and network interface) on a conventional computer. When it boots on a conventional computer, the operating system uses a hardware mechanism known as a *bus* to communicate with I/O devices. The first step consists of sending a series of requests across the bus to form a list of all I/O devices that are present. The operating system must include *device driver* software for each device. The operating system uses the device driver code for a given device to control the device hardware and handle all communication with the device.

We can now understand why a dilemma arises when a server runs multiple virtual machines. Each virtual machine will run its own operating system, and a virtual machine can boot a standard operating system. On the one hand, a standard operating system will try to use the bus to take ownership of all I/O devices. On the other hand, the hypervisor cannot allow one of the operating systems to gain exclusive control of I/O devices because the devices must be usable by all virtual machines. In particular, because all VMs need network communication, no single VM can be allowed to commandeer the server's network I/O.

5.9 Virtual I/O Devices

VM technology uses an interesting approach to handle I/O: *virtual I/O devices*. When it creates a VM, a hypervisor creates a set of *virtual I/O devices* for the VM to use. A virtual I/O device is implemented by software. When an operating system on a VM attempts to use the bus to access an I/O device, the access violates privilege, which means the hypervisor is invoked. The hypervisor runs the appropriate virtual device software, and then arranges to send the response to the operating system as if a physical device responded. That is, the hypervisor makes it appear that the response appears to come over the bus. The point is:

> *From an operating system's point of view, a virtual device is indistinguishable from a real device because communication with a virtual device uses the same bus operations that communication with a physical device uses.*

5.10 Virtual Device Details

The use of virtual devices introduces two approaches, and both have been used:

- Invent a new, imaginary device
- Emulate an existing hardware device

Invent a new, imaginary device. Because it is implemented with software, a virtual device can have any properties a programmer imagines†. Communication between the operating system and the device uses the standard bus operations, but the details of what I/O operations the device supports and how data moves between the operating system and the device are controlled by the virtual device, which gives an opportunity for an improved design. Most commercial device hardware uses a complicated, arcane interface that makes writing device driver software difficult. In principle, a programmer who creates a virtual device can create a beautiful, clean design that avoids the messy hardware details, making it much easier to write device driver software. It may also be possible to increase efficiency (e.g., by creating an imaginary disk with large blocks).

Emulate an existing hardware device. The alternative to creating an imaginary device consists of building software that emulates an existing hardware device. To understand the motivation for emulation, consider the availability of device driver software. When it boots and discovers a device, an operating system must load device driver software that can interact with the device (i.e., send I/O requests to the device and receive responses). Because the device driver software used with one operating system is incompatible with other operating systems, a separate driver must be created for each operating system. Writing and debugging device driver code is tedious, time-consuming, and requires expertise with a specific operating system. Building virtual device software that emulates an existing hardware device avoids having to build new device drivers —an operating system can load and run existing device driver software.

The downside of building a virtual device that emulates an existing hardware device arises because the virtual device must behave exactly like a real hardware device. It must accept all the same requests and respond to each request exactly like the hardware device responds. Nevertheless, virtual machine systems have been deployed in which virtual devices emulate commercial hardware devices.

---

†For an example, see *virtIO* at URL https://docs.oasis-open.org/virtio/virtio/v1.0/virtio-v1.0.html.

5.11 An Example Virtual Device

As an example of a virtual device, consider a *virtual disk*†. Because data centers locate storage facilities separate from servers, all disk I/O requires communication over the data center network. Virtual disk software handles the situation easily by providing a standard disk interface to an operating system and also communicating over the network. Figure 5.4 illustrates the conceptual arrangement.

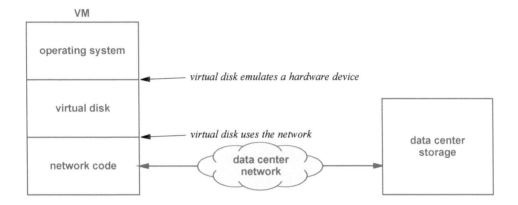

Figure 5.4 The conceptual organization of an operating system, virtual disk code, and network code.

As the figure shows, virtual disk code occupies a position between an operating system that runs in a VM and the network code in the hypervisor. When it communicates with the operating system, the virtual disk acts like a hardware device, using standard bus communication to receive requests and supply responses. For a disk, a request either writes data to the disk or reads data from the disk. Each time it receives a request, the virtual disk code uses the network to communicate with the storage facility in the data center. The virtual disk specifies the VM that made a request, and either sends data to be stored on the VM's disk or requests a copy of data from the VM's disk.

5.12 A VM As A Digital Object

A VM differs from a real server in a significant way: a VM is created and managed entirely by software. A hypervisor must keep a record of the VM, the region(s) of memory that have been allocated to the VM, the virtual I/O devices that have been created for the VM (including disk space that has been allocated in the data center storage facility), and the current status of the VM (e.g., whether the VM is currently running or has been suspended to allow another VM to run).

---

†Chapter 8 describes virtual disks and other data center storage mechanisms in more detail.

Having a complete record of each VM in a hypervisor has an interesting consequence: a VM can be turned into a digital object. That is, the entire VM can be
transformed into a set of bytes. To understand how, imagine that a VM is currently
suspended. The hypervisor has a record of all the memory the VM is using, so the
VM's memory segments can be collected. Imagine, for example, that they are placed in
a special file. The code and data for the operating system and apps that the VM is running are all stored in memory, so when the VM's memory has been collected, the OS
and apps will be collected. Similarly, the virtual devices that have been created for the
VM consist of software, so they can be collected as well. The key idea is:

> *Because a VM is implemented with software, all the pieces of a VM*
> *can be collected together into a digital object.*

5.13 VM Migration

The ability to collect all the pieces of a VM into a digital object has great signifi-
cance, and data centers take advantage of it. Suppose, for example, that a data center
manager receives an alert that a power supply is about to fail on a server. The manager
can instruct the hypervisor to stop all the VMs running on the server, save a copy of
each VM on disk, and then shut the server down. The manager can replace the server
hardware, reboot the server, reload all the VMs, and resume operating, allowing each of
them to continue where it left off.

Although it allows one to stop and restart a server, the ability to convert a VM to a
digital object enables something much more important in cloud data centers: *VM migration*. That is, a VM can be moved. The basic idea is straightforward: stop a VM that is
running on one server, convert the VM to a digital object, send the bytes across the network to a new server, and resume the VM on the new server. That is, a hypervisor on
one server converts a VM into a digital object, sends the object to the hypervisor on
another server, and the receiving hypervisor resumes the VM on its server. Although
the VM is moved and the hypervisors do the work, the cloud industry says the VM *mi-
grates* from one server to another.

The ability to migrate VMs allows a provider to rebalance loads dynamically to
eliminate hot spots. Suppose, for example, that the VMs running on server X all perform intensive long-running computations with little or no I/O activity. Further suppose
that the VMs running on server Y transfer huge amounts of data, but do not spend much
time computing. The processor on server X will be saturated while the processor on
server Y will remain relatively idle. Meanwhile, the I/O bus on server Y will be saturated while the bus on X is relatively idle. Migration allows a data center owner to
balance the two servers by moving some of the VMs from Y to X and some of the VMs
originally on X to Y. As a result, neither server will have a saturated processor or a saturated I/O bus.

Migration has many other uses. For example, during periods of especially light load, a data center manager can migrate VMs away from an area of servers and then power down the empty servers to reduce power consumption. A set of VMs owned by a given customer can be migrated to the same pod (or adjacent pods), reducing the network traffic that must cross the data center and reducing the latency among the customer's VMs.

5.14 Live Migration Using Three Phases

We have seen that it is possible to convert a VM into a digital object if the VM has been stopped. Unfortunately, it takes time to stop a VM, send it across a network, and then restart it. What happens if the VM is communicating over the network? Perhaps the VM is accessing a database or downloading a file. In such cases, stopping a VM can cause the other side of the communication to give up or declare the VM to have aborted. Instead of stopping a VM, engineers have devised a migration scheme that allows *live migration*. That is, a VM continues to run while migration proceeds.

To enable live migration, the amount of time a VM is unavailable must be minimized. The technique used divides migration into three phases:

- Phase 1: pre-copy
- Phase 2: stop-and-copy
- Phase 3: post-copy

Phase 1: pre-copy. In phase 1, the entire memory of the VM is copied to the new server while the VM continues to run. Of course, some pages in memory will be changed after they are copied; a record is kept of the pages that changed.

Phase 2: stop-and-copy. In phase 2, the VM is temporarily suspended, and any pages that changed after the phase 1 copy are copied again. The virtual memory system on modern servers makes it easy to detect which pages have been changed (i.e., are *dirty*). Thus, the set of changed pages can be identified quickly.

Phase 3: post-copy. In phase 3, the hypervisor sends remaining state information to the hypervisor on the new server. The state information includes items that are needed to resume (i.e., unsuspend) the VM, including the contents of registers when the VM was suspended. The hypervisor on the new server uses the information to allow the VM to continue executing.

Although a VM must be suspended temporarily, pre-copying the entire memory means almost no time is required during the second and third phases. To summarize:

A three-phase approach that minimizes the time a VM is unavailable makes live migration feasible.

5.15 Running Virtual Machines In An Application

The previous sections describe how VM technology uses a hypervisor that runs directly on server hardware. Interestingly, an alternative form of VM technology has been developed that allows a hypervisor to run on a conventional operating system (i.e., on a user's laptop). A user launches a *hypervisor application* called a *hosted hypervisor*†. A hosted hypervisor runs along with other applications. Once it starts, the hosted hypervisor allows a user to create and manipulate VMs.

On a conventional computer, we think of an operating system rather than a hypervisor as owning and controlling the hardware. We use the term *host operating system* to describe the main operating system. When the user launches an application, the application runs as an independent process. Similarly, when the user launches a hosted hypervisor, the hypervisor runs as an application process. We say that the hosted hypervisor runs in *user space*.

As expected, each VM must run an operating system, which is called a *guest operating system*. A guest operating system does not need to be the same as a host operating system, and guest operating systems may differ from one another. For example, a user with a Mac laptop can use a hosted hypervisor to create a VM that runs Linux and another VM that runs Windows. Figure 5.5 shows the arrangement of software components after a user has launched a hosted hypervisor that runs two VMs along with four other applications.

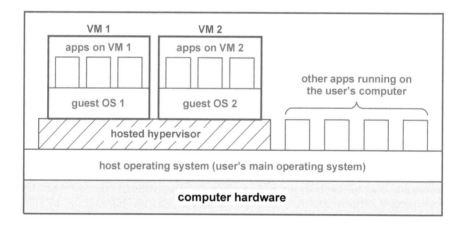

Figure 5.5 The conceptual arrangement of software on a conventional computer that results when a user runs a hypervisor application and creates two VMs.

Three questions arise concerning a system that runs a hosted hypervisor on a conventional computer.

---

†A hosted hypervisor is also classified as a *type 2* hypervisor to distinguish it from a *type 1 hypervisor* that runs directly on server hardware.

- How is it possible?
- What benefit does it offer a user?
- Is the technology useful in a cloud data center?

Once we understand how a hosted hypervisor works, we will discuss how a user can benefit. The next chapter answers the question about whether the technology is useful in a cloud data center by describing one possible use: running a guest operating system that can run containers.

5.16 Facilities That Make A Hosted Hypervisor Possible

Two facilities allow a hosted hypervisor and a guest OS to operate correctly without high privilege: a processor mechanism that supports virtualization and a way to map guest OS operations onto the host operating system's services. Modern processors include a special virtualization mechanism that enables a hosted hypervisor to create VMs that each run a guest OS. Although the mechanism involves many complex details and subtleties, the overall result is straightforward: the mechanism allows a guest OS to proceed as if it were running at the highest privilege level, even though it does not have privilege and does not have complete access to the underlying hardware. For example, although it is only given access to a portion of the physical memory, a guest OS remains unaware that it has a limited view.

A hosted hypervisor arranges a mapping between I/O requests that a guest OS makes and services in the host operating system. For example, when a guest OS communicates over a network, the hosted hypervisor forwards the requests to the host operating system in such a way that the guest OS has the illusion of using a hardware interface. The hypervisor may choose to assign each guest OS a unique network address, making it appear that each guest OS has its own network interface. Similarly, the hypervisor forwards disk I/O requests to the host operating system in such a way that it gives the guest OS the illusion of a separate disk.

Interestingly, a hosted hypervisor can arrange for some files or file systems to be shared between the guest and host operating systems. Shared files provide an efficient way for applications running in a VM to share data with applications running on the host. The point is:

> *Because a hosted hypervisor can map I/O requests from a guest operating system onto services offered by the host operating system, it is possible for applications running on VMs and applications running on the host to share files.*

5.17 How A User Benefits From A Hosted Hypervisor

A hosted hypervisor allows a user to run multiple operating systems on a single computer. Multiboot mechanisms also allow a user to boot multiple operating systems, but a user can only boot one system at a given time. A hosted hypervisor has a distinct advantage because a user can run multiple systems simultaneously and can switch from one system to another without rebooting the computer.

To understand how such a system can be useful, suppose a user who runs Mac OS on their laptop wishes to use an application that only runs on Linux. With a hosted hypervisor, the user can launch a hosted hypervisor, boot Linux in a window (or virtual desktop), and then use Linux to run the desired application. Furthermore, the user can leave the virtual machine running, and switch back and forth between Linux and Mac OS simply by switching windows or desktops. The point is:

> *Unlike a multiboot mechanism, a hosted hypervisor allows a user to run multiple operating systems simultaneously and switch among them quickly.*

5.18 Summary

Cloud data centers make extensive use of Virtual Machine (VM) technology. Software known as a hypervisor allows a user to create and manage VMs. Conceptually, the hypervisor owns the underlying hardware, VMs run over the hypervisor, and each VM runs an operating system plus apps. In practice, the code for a hypervisor, operating system, and apps reside in memory, and the hardware executes code directly until the code attempts to perform an operation that requires higher privilege. If an app attempts to exceed its privilege level, control passes to the operating system that started the app. If an operating system attempts to exceed its privilege level, control passes to the hypervisor. We say that the levels of privilege impose a hierarchy of trust.

A hypervisor provides each operating system with the illusion that it has direct access to peripheral devices over a conventional I/O bus. The mapping allows for arbitrary action, such as sending requests over a network to a remote storage facility when an operating system performs disk I/O.

All pieces of a VM can be collected into a digital object, which can be sent over a network, allowing hypervisors to migrate a VM from one physical server to another. A three-phase migration procedure minimizes the time during which a VM is unavailable, making live migration feasible. Data center owners use migration to rebalance the load on servers.

An alternative form of VM technology allows a user to run a hosted hypervisor on a conventional computer. Doing so means a user can switch among operating systems without rebooting the computer.

Chapter Contents

6

Containers

6.1 Introduction

The previous chapter examines one form of server virtualization: virtual machines (VMs). The chapter states that VMs used in data centers emulate commercial hardware, and explains why an app running in a VM can execute at hardware speed. The chapter discusses how a VM can be suspended, converted to an object, and migrated from one physical server to another, and describes how a data center manager can use migration.

This chapter considers an alternative form of server virtualization that has become popular for use in data centers: container technology. The chapter explains the motivation for containers and describes the underlying technology.

6.2 The Advantages And Disadvantages Of VMs

To understand the motivation for containers, we must consider the advantages and disadvantages of VMs. The chief advantage of the VM approach lies in its support of arbitrary operating systems. VM technology virtualizes processor hardware and creates an emulation so close to an actual processor that a conventional operating system built to run directly on hardware can run inside a VM with no change. Once a hypervisor has been loaded onto a server, the hypervisor can create multiple VMs and arrange for each to run an operating system that differs from the operating systems running in other VMs. Once an operating system has been started in a VM, apps built to use the operating system can run unchanged. Thus, a cloud customer who leases a VM has the freedom to choose all the software that runs in the VM, including the operating system. Chapter 5 describes techniques that allow both the operating system code and apps running in a VM to execute at hardware speed.

VM technology also has some disadvantages. Creating a VM takes time. In particular, VM creation requires booting an operating system. Furthermore, the VM approach places computational overhead on the server. Consider, for example, a server that runs four VMs, each with its own operating system. Figure 6.1 illustrates the configuration.

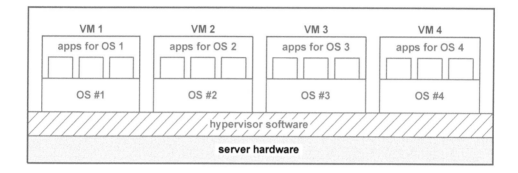

Figure 6.1 Illustration of four VMs each running their own operating system.

In the figure, four operating systems are running on the server, and each injects two forms of overhead. First, each operating system runs a scheduler that switches the processor among the apps it runs. Instead of the overhead from one scheduler, a server running VMs experiences scheduling overhead from each of the four. Second, and most important, operating systems run additional background processes to handle tasks for apps, such as processing network packets. Instead of one set of background processes, a server that runs VMs will run background processes for each VM. As additional VMs are added to a server, the overhead increases. We can summarize:

> *Although VM technology has the advantage of allowing the owner of a VM to choose an operating system, VM creation entails the overhead of booting an operating system, and running multiple VMs on a server imposes computational overhead because each operating system schedules apps and runs background processes.*

6.3 Traditional Apps And Elasticity On Demand

VM technology works well in situations where a virtual server persists for a long time (e.g., days) or a user needs the freedom to choose an operating system. In many instances, however, a user only runs a single application and does not need all the facilities in an operating system. A user who only runs a single app can use a cloud service to handle rapid elasticity —the number of copies of the app can increase or decrease quickly as demand rises and falls. In such cases, the overhead of booting an operating system makes VM technology unattractive. The question arises: "Can an alternative virtualization technology be devised that avoids the overhead of booting an operating system?"

A conventional operating system includes a facility that satisfies most of the need: support for concurrent processes. When a user launches an app, the operating system creates a process to run the app, and process creation takes much less time than booting an operating system. When a user decides to shut down an app, the operating system can terminate a process quickly. One can imagine a cloud system that allows customers to deploy and terminate copies of an app. Such a facility would consist of physical servers each running a specific operating system. When a customer requests an additional copy of a particular app, the provider uses software to choose a lightly-loaded server and launch a copy of the app. Similarly, when a customer requests fewer copies, the provider uses software to locate and terminate idle copies.

Unfortunately, apps running on an operating system do not solve the problem completely because an operating system does not ensure complete isolation among apps run by multiple tenants. In particular, most operating systems assume apps will share network access. That is, an operating system obtains an Internet address and allows all apps to use the address. An operating system usually provides a file system that all apps share. In addition to sharing, most operating systems allow a process to obtain a list of other processes and the resources they have consumed, which means a tenant might be able to deduce what other tenants' apps are doing. The point is:

> *Although it allows apps to be started and terminated quickly, a traditional operating system running apps does not suffice for a multitenant cloud service because processes share facilities, such as a network address and a file system that allow an app to obtain information about other apps.*

6.4 Isolation Facilities In An Operating System

Starting with the earliest systems, operating systems designers have discovered ways to isolate the computations and data owned by one user from those of another. Most operating systems use virtual memory hardware to provide each process with a separate memory address space and ensure that a running app cannot see or alter memory locations owned by other apps. User IDs provide additional isolation by assigning an owner to each running process and each file, and forbidding a process owned by one user from accessing or removing an object owned by another user. However, the user ID mechanism in most operating systems does not scale to handle a cloud service with arbitrary numbers of customers. Therefore, the operating systems community has continued to investigate other ways to isolate running apps.

6.5 Linux Namespaces Used For Isolation

Some of the most significant advances in isolation mechanisms have arisen in the open source community; cloud computing has spurred adoption. Under various names, such as *jails*, the community has incorporated isolation mechanisms into the Linux operating system. Known as *namespaces*, a current set of mechanisms can be used to isolate various aspects of an application. Figure 6.2 lists seven major namespaces used with containers.

Namespace	Controls
Mount	File system mount points
Process ID	Process identifiers
Network	Visible network interfaces
Interprocess communication	Process-to-process communication
User ID	User and group identifiers
UTS	Host and domain names
Control group	A group of processes

Figure 6.2 The seven major namespaces in the Linux operating system that can be used to provide various forms of isolation.

The mechanisms control various aspects of isolation. For example, the process ID namespace allows each isolated app to use its own set of process IDs 0, 1, 2..., and so on. Instead of all processes sharing the single Internet address that the host owns, the network namespace allows a process to be assigned a unique Internet address. More important, the application can be on a different virtual network than the host operating system. Similarly, instead of all processes sharing the host's file system, the mount namespace allows a process to be assigned a separate, isolated file namespace (similar to the *chroot* mechanism introduced in earlier Unix systems). Finally, a control group namespace allows one to control a group of processes.

One of the most interesting aspects of namespace isolation arises because the mechanisms block access to administrative information about other isolated applications running on the same host. For example, the process ID namespace guarantees that if an isolated application X attempts to extract information about a process with ID p, the results will only refer to processes in X's isolated environment, even if one or more of the other isolated apps running on the same host happen to have a process with ID p. The important idea is:

Isolation facilities in an operating system make it possible to run multiple apps on a computer without interference and without allowing an app to learn about other isolated apps.

6.6 The Container Approach For Isolated Apps

When an app uses operating system mechanisms to enforce isolation, the app remains protected from other apps. Conceptually, we think of the app as running in its own environment, surrounded by walls that keep others out. Industry uses the term *container* to capture the idea of an environment that surrounds and protects an app while the app runs. At any time, a server computer runs a set of containers, each of which contains an app. Figure 6.3 illustrates the conceptual organization of software on a server that runs containers.

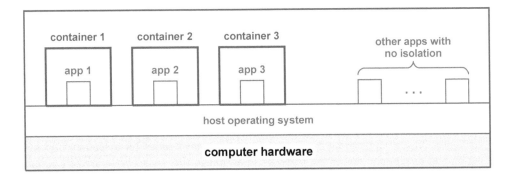

Figure 6.3 The conceptual organization of software when a server computer runs a set of containers.

As the figure illustrates, it is possible to run conventional apps outside of containers at the same time as containers. Although conventional, unprivileged apps cannot interfere with apps in containers, a conventional app may be able to obtain some information about the processes running in containers. Therefore, systems that run containers usually restrict conventional apps to the control software used to create and manage containers.

We can summarize:

> *A container consists of an isolated environment in which an application can run. A container runs on a conventional operating system, and multiple containers can run on the same operating system concurrently. Each container provides isolation, which means an application in one container cannot interfere with an application in another container.*

6.7 Docker Containers

One particular container technology has become popular for use with cloud systems. The technology resulted from an open source project known as *Docker†*. The Docker approach has become prominent for four main reasons:

- Tools that enable rapid and easy development of containers
- An extensive registry of software for use with containers
- Techniques that allow rapid instantiation of an isolated app
- Reproducible execution across hosts

Development tools. The Docker technology provides an easy way to develop apps that can be deployed in an isolated environment. Unlike a conventional programming system in which a programmer must write significant pieces of code, Docker uses a high-level approach that allows a programmer to combine large pre-built code modules into a single image that runs when a container is deployed.

The Docker model does not separate a container from its contents. That is, one does not first start a Docker container running and then choose an app to run inside the container. Instead, a programmer creates all the software needed for a container, including an app to run, and places the software in an image file. A separate image file must be created for each app. When an image file runs, a container is created to run the app. We say the app has been *containerized*.

Extensive registry of software. In addition to the basic tools programmers can use to create software for containers, the Docker project has produced *Docker Hub*, an extensive registry of open source software that is ready to use. The registry enables a programmer to share deployable apps, such as a web server, without writing code. More important, a user (or an operator) can combine pieces from the registry in the same way that a conventional program uses modules from a library.

Rapid instantiation. Because a container does not require a full operating system, a container is much smaller than a VM. Consequently, the time required to download a container can be an order of magnitude less than the time required to download a VM. In addition, Docker uses an early binding approach that combines all the libraries and other run-time software that will be needed to run the container into an image file. Unlike a VM that must wait for an operating system to boot, a container can start instantly. Furthermore, unlike a conventional app that may require an operating system to load one or more libraries dynamically, the early binding approach means a Docker container does not need the operating system to perform extra work when the container starts. As a result, the time required to create a container is impressively small.

Reproducible execution. Once a Docker container has been built, the container image becomes *immutable* —the image remains unchanged, independent of the number of times the image runs in a container. Furthermore, because all the necessary software components have been built in, a container image performs the same on any system. As a result, container execution always gives reproducible results. The point is:

---

†As a follow-on to the Docker project, a company named *Docker Incorporated* was created to build and sell orchestration software that deploys and manages Docker containers.

Docker technology makes app development easy, offers a registry of pre-built software, optimizes the time required to start a container, and guarantees reproducible execution.

6.8 Docker Terminology And Development Tools

Like many large efforts, the Docker project has spawned new terminology. In addition to tools used to create and launch an app, the extended environment includes tools used to deploy and control copies of running containers. Rather than examine all aspects of Docker, we will focus on the basics. The table in Figure 6.4 lists a few terms that Docker uses, along with items from conventional computing systems that provide an analogy to help clarify the concept.

Docker Term	Meaning	Analogous To
image	A file that contains binary code for a container along with its dependencies	a.out file (MacOS/Linux) .exe file (Windows)
layer	One piece of software added to an image as it is built	a software module
container	An instance of an image	an executing process
Dockerfile	A specification for how to build an image	Makefile (Linux)
docker build	A command that constructs an image according to a Dockerfile	"make" program (Linux)
docker run X	A command that runs image X as a container	launch app X

Figure 6.4 Basic terms used by Docker and their meaning.

A Docker *image* is a file that can be executed, and a *container* is the execution of an image. Some ambiguity occurs because an image can have two forms: a *partial image* that forms a building block, and a *container image* that includes all the software needed for a particular app. Partial images form a useful part of the Docker programming environment analogous to library functions in a conventional programming environment. That is, Docker offers a registry of pre-built partial images that each handle a common task.

Docker uses the term *layer* to refer to each piece of code a programmer includes in a container image. We say that the programmer starts by specifying a *base image* and then adds one or more layers of software. For each layer, a programmer has a choice:

the programmer can write code from scratch or download one of the pre-built building blocks from a registry.

Docker does not use conventional compiler tools to build an image. Instead, it uses a special *build* facility analogous to Linux's *make*. A programmer creates a text file with items that specify how to construct a container image. By default Docker expects the specifications to be placed in a text file named *Dockerfile*. A programmer runs *docker build* which reads *Dockerfile*, follows the specified steps, constructs a container image, and issues a log of steps taken (along with error messages, if any).

A Docker container image is not an executable file, and cannot be launched the same way one launches a conventional app. Instead, one must use the command *docker run* to specify that an image is to be run as a Docker container. The next section describes the structure of Docker software, and explains why containers require a non-standard approach.

6.9 Docker Software Components

Although Figure 6.3† illustrates a set of containers running independently the same way a process runs, execution of Docker containers requires additional software support. In particular, containers operate under the control of an application known as a *Docker daemon* (*dockerd*). In addition, Docker provides a user interface program, *docker*. Collectively, the software is known as the *Docker Engine*. Figure 6.5 illustrates Docker software and shows that dockerd manages both images that have been saved locally and running containers.

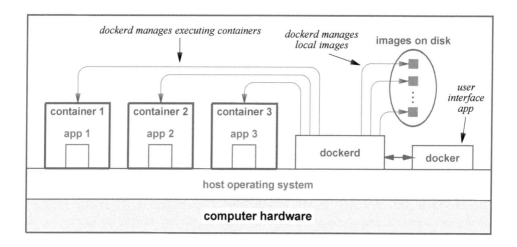

Figure 6.5 Illustration of the Docker daemon, *dockerd*, which manages both containers and images.

---

†Figure 6.3 can be found on page 75.

The dockerd program, which remains running in background at all times, contains several key subsystems. In addition to a subsystem that launches and terminates containers, dockerd contains a subsystem used to build images and a subsystem used to download items from a registry.

A user does not interact with dockerd directly. Instead, dockerd provides two interfaces through which a user can make requests and obtain information:

- A RESTful interface intended for applications

- A Command Line Interface (CLI) intended for humans

RESTful Interface. As the next section of the text explains, customers of cloud services do not usually create and manage containers manually. Instead, they use *orchestration software* to deploy and manage sets of containers. When it needs to create, terminate, or otherwise manage containers, orchestration software uses dockerd's RESTful interface. As expected, the RESTful interface uses the HTTP protocol, and allows orchestration software to send requests and receive responses.

Command Line Interface. To accommodate situations when a human needs to manage containers manually, dockerd offers an interactive command-line interface that allows a user to enter one command at a time and receive a response. When a user installs Docker software, the installation includes an application named *docker* that provides the CLI. To send a command to dockerd, a user can type the following in a terminal window†:

docker command arguments...

The Docker software provides an extensive set of commands that include fifteen management commands (e.g., to manage containers) plus forty additional commands. Rather than attempt to describe all possible commands, we will examine a few commands that help illustrate some of the functionality. Figure 6.6 lists example commands.

Command	Meaning
docker build .	Read Dockerfile in the current directory and follow the directives to construct an image
docker images	List all images on the local machine
docker run	Create and start a container running a specified image
docker ps	List all currently running containers on the system
docker pull	Download a copy of an image from a registry
docker stop	Stop one or more running containers
docker start	Restart one or more stopped containers

Figure 6.6 Examples of Docker commands a user can enter through the CLI.

---

†In fact, it is possible to run the docker command on one computer and connect to dockerd on another computer.

To construct a container image, a programmer creates a *Dockerfile*, places the file in the current directory, and runs:

```
docker build .
```

where the dot specifies that Docker should look in the current directory for a Dockerfile that specifies how to build the image.

Recall that Docker does not store images in the user's directory. Instead, dockerd stores the images. When it completes successfully, a build command will print a name for the container image that was produced and stored. An image name consists of a hash that is twelve characters long and meaningless to a human. For example, Docker might use the hash *f34cd9527ae6* as the name of an image.

To run an image as a container, a programmer invokes the *run* command and supplies the image name. For example, to run the image described above, a programmer enters:

```
docker run f34cd9527ae6
```

Docker stores images until the user decides to remove them. To review the list of all saved images, a user can enter:

```
docker images
```

which will print a list of all the saved images along with their names and the time at which each was created.

6.10 Base Operating System And Files

Recall that a container uses facilities in the host operating system. However, in addition to layers of app software, most container images include a layer of software known as a *base operating system* that acts as an interface to the underlying host operating system. Figure 6.7 illustrates the idea.

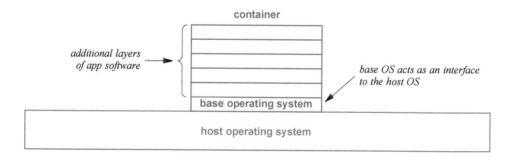

Figure 6.7 Illustration of a small base operating system in a container.

Think of a base operating system as a shim layer between app software in the container and the underlying host operating system. Apps running in the container make calls to the base operating system which then makes calls to the host operating system. When building a container image, a programmer starts by specifying a base operating system; base operating systems are available in registries.

From a programmer's perspective, most of the distinctions between a base operating system and the host operating system are unimportant. A programmer can use the base operating system to run commands or launch apps as usual. The apps make operating system requests as usual (e.g., to communicate over a network). If a user runs an interactive app in a container (e.g., a shell) and runs the container interactively, the app will connect to the user's standard input and output devices, just as if the programmer launched a conventional app. For example, when a user runs a container interactively in a terminal window and an app running the container writes to standard output, the output will appear in the user's terminal window and keystrokes entered in the window will be sent to the app running in the container.

One aspect of containers differs from conventional apps: the file system. When a container image is created, the base operating system includes a small file system that holds apps and other data. Apps running in the container can read files and can create new files. Because a container is immutable, however, files created when a container runs will not persist across invocations of the container image —each time a user creates a container to run a given image, the file system in the container will start with the files that were present when the image was initially created.

Chapter 8 explains that when a container runs, a user has the option of connecting directories in the container to persistent files, either in the file system of the host operating system or on data center storage. For now, it is sufficient to know that unless steps are taken, files created in the container will not persist once the container terminates. We can summarize:

> *Unless a programmer connects a container to permanent storage, changes made to local files when a container runs will not be saved for subsequent invocations of the container image.*

6.11 Items In A Dockerfile

A Dockerfile specifies a sequence of steps to be taken to build a container image. This section provides a short overview of the items that can appear in a Dockerfile†.

Each item in a Dockerfile begins with a keyword. The following paragraphs each describe one of the keywords.

FROM. The *FROM* keyword, which specifies a base operating system to use for the image, must appear as the first keyword in a Dockerfile. For example, to use the alpine Linux base, one specifies:

```
FROM alpine
```

---

†The items described in this chapter have been taken from Docker release 19.03.5; for more details, see https://docs.docker.com/engine/reference/builder/.

When it builds an image, Docker obtains and uses the latest version of the base system that is available. To specify an older version, a programmer can add a colon and a version number to the name. For example, alpine:3.5 specifies version 3.5 of alpine.

RUN. The *RUN* keyword specifies that a program should be run to add a new layer to the image. The name is unfortunately confusing because the "running" takes place when the image is built rather than when the container runs. For example, to execute the *apk* program during the build and request that it add Python and Pip to the image, a programmer uses:

```
RUN apk add py2-pip
```

ENTRYPOINT. A programmer must specify where execution begins when a container starts running. To do so, a programmer uses the *ENTRYPOINT* keyword followed by the name of an executable program in the image file system and arguments that will be passed to the program. ENTRYPOINT has two forms: one in which arguments are passed as literal strings and one that uses a shell to invoke the initial program. Using a shell allows variable substitution, but the non-shell option is preferred because the program will run with process ID 1 and will receive Unix signals.

CMD. The CMD keyword, which is related to ENTRYPOINT, has multiple forms and multiple purposes. As a minor use, CMD can be used as an alternative to ENTRYPOINT to specify a command to be run by a shell when the container starts; a programmer cannot specify an initial command with both CMD and ENTRYPOINT. The main purpose of CMD focuses on providing a set of default arguments for ENTRYPOINT that can be overridden when the container is started.

COPY and *ADD.* Both the *COPY* and the older *ADD* keywords can be used to add directories and files to the file system being constructed for the image (i.e., the file system that will be available when the image runs in a container). COPY is straightforward because it copies files from the computer where the image is being built into the image file system. As an example, a programmer might build a Python script, test the script on the local computer, and then specify that the script should be copied into an image.

The ADD keyword allows a programmer to specify local files or give a URL that specifies a remote file. Furthermore, ADD understands how to decompress and open archives. In many cases, however, opening an archive results in extra, unneeded files in the image, making the image larger than necessary. Therefore, the recommended approach has become: before building an image, download remote files to the local computer, open archives, remove unneeded files, and use the COPY keyword to copy the files into the image.

EXPOSE and *VOLUME.* Although EXPOSE deals with Internet access and VOLUME deals with file systems, both specify a way to connect a container to the outside world. EXPOSE specifies protocol port numbers that the container is designed to use. For example, an image that contains a web server might specify that the container is designed to use port 80:

```
EXPOSE 80
```

VOLUME specifies a mount point in the image file system where an external file system can connect. That is, VOLUME provides a way for a container to connect to persistent storage in a data center. VOLUME does not specify a remote storage location, nor does it specify how the connection to a remote location should be made. Instead, such details must be specified when a user starts a container.

Figure 6.8 summarizes:

Item	Meaning
FROM	Specifies a base operating system for the image
RUN	Runs a program to fetch an item for the image
ENTRYPOINT	Specifies an app that runs when the container starts
CMD	Specifies a command that runs when the container starts
COPY	Copies local files into the image file system
ADD	An earlier form of COPY sometimes still seen
EXPOSE	Specifies network port numbers that the container will use
VOLUME	Specifies mount points for an external file system

Figure 6.8 Examples of keywords used for items in a Dockerfile.

6.12 An Example Dockerfile

Figure 6.9 contains a trivial example of a Dockerfile that specifies *alpine* to be the base operating system and specifies that file */bin/echo* (i.e., the Linux *echo* command) should be run when the container starts. When it runs as a container, the image prints *hi there*†. Once it runs, the echo command exits, which causes the container to exit.

```
FROM alpine
ENTRYPOINT ["/bin/echo", "hi there"]
```

Figure 6.9 A trivial example of a Dockerfile that specifies running the *echo* command.

6.13 Summary

Container technology provides an alternate form of virtualization that avoids the overhead incurred when a guest operating system runs in each VM. Although it operates much like a conventional process, a container uses namespace mechanisms in the host operating system to remain isolated from other containers.

---

†The example uses "hi there" instead of the classic "hello world" message to avoid being confused with the *hello-world* image available with Docker.

A popular container technology arose from the open source Docker project. To build a container, a programmer writes specifications in a text file (Dockerfile). Specifications include a base operating system to use, other software that should be added to the container, an initial program to execute when the container starts, and external network and file system connections that can be used when an image runs in a container. To construct a container image, a programmer invokes Docker's *build* mechanism, which follows the specifications and produces an image. Once it has been created, an image can be run in a container.

Docker software includes two main pieces: a daemon named *dockerd* that runs in background, and a command named *docker* that a user invokes to interact with dockerd through a command-line interface. The docker interface allows a user to build an image or run and manage containers. Images are not stored in the user's directory; dockerd stores images and controls running containers. Dockerd also offers a RESTful interface that allows orchestration software to manage large-scale deployments of containers.

Chapter Contents

7

Virtual Networks

7.1 Introduction

Chapter 4 describes networks used in data centers. The chapter defines the east-west traffic pattern, and explains that a leaf-spine architecture that handles such traffic has evolved to replace the older hierarchical architecture. The chapter also points out that instead of requiring a data center to add new switches with higher capacity links, the leaf-spine approach accommodates scaling merely by adding additional switches.

This chapter explores the complex topic of network virtualization. Instead of attempting to cover all technologies or presenting low-level technical details, the chapter considers the motivation for virtualized networks, the general concepts, the use of SDN, and the ways cloud providers can employ virtualization technologies.

7.2 Conflicting Goals For A Data Center Network

Interestingly, data center network designers face a challenge of balancing between two conflicting goals:

- Universal connectivity
- Safe, isolated communication

Universal connectivity. It should be obvious that systems across the entire data center must be able to communicate with one another. When they run, apps communicate with databases, the apps running on other servers, storage facilities, and, possibly, with computers on the global Internet. Furthermore, to give a provider freedom to place

VMs and containers on arbitrary physical servers, the network must guarantee *universal connectivity* —a pair of apps must be able to communicate independent of the physical servers on which they run. Chapter 4 describes the leaf-spine network architecture in which each server connects to a leaf switch (i.e., a Top-of-Rack switch), and the leaf switches connect to spines. Clearly, the architecture provides universal connectivity. In fact, multiple physical paths exist between any two servers.

Safe, isolated communication. Although universal connectivity across a data center is required, a customer of a multi-tenant data center also demands a guarantee that outsiders will not have access to the customer's systems or communication. That is, a customer needs assurance that their computing systems are isolated from other tenants and their communication is safe. Ideally, each customer would like a separate network that only connects the customer's VMs and containers. To summarize the situation:

> *On the one hand, to permit VMs and containers to communicate between arbitrary physical servers, a data center network must provide universal connectivity. On the other hand, a given tenant seeks a network architecture that keeps their VMs and containers isolated and safe.*

7.3 Virtual Networks, Overlays, And Underlays

How can a provider keep each tenant's traffic isolated from other tenants' traffic? A cloud service with thousands of tenants makes separate physical networks impractical. The answer lies in an approach known as *network virtualization*. The basic idea is straightforward: build a network with universal connectivity, and then configure the network switches to act as if they control a set of independent networks. That is, a provider imagines a set of independent networks, and configures each switch to keep traffic sent across one of the imagined networks separate from traffic on other imagined networks. To emphasize that each of the imagined networks is only achieved by restricting the set of VMs and containers that receive a given packet, we say that the configuration creates a set of *virtual networks*†.

Conceptually, each virtual network links a tenant's virtual machines and containers. Figure 7.1 illustrates the idea by showing two virtual networks that each link four VMs. Keep in mind that the two networks in the figure are fiction. In reality, each VM runs on a server in a rack, and racks are connected by spine switches. Suppose, for example, that VM 2 and VM 5 run on servers in separate racks. The network path between them includes two leaf switches and at least a spine switch.

We use the term *overlay network* to refer to a virtual network that does not actually exist but which in effect has been created by configuring switches to restrict communication. We use the term *underlay network* to refer to the underlying physical network that provides connections among entities in a virtual network.

---

†The term *logical network* is synonymous with *virtual network*.

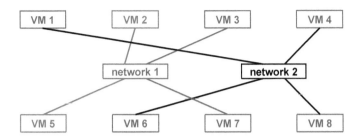

Figure 7.1 Illustration of two virtual networks that each connect four VMs. Each virtual network forms an *overlay*.

We can summarize the terminology.

> *The term* overlay network *refers to a virtual network created by configuring switches, and the term* underlay network *refers to the underlying physical network that provides actual network connections.*

7.4 Virtual Local Area Networks (VLANs)

Cloud providers have used a variety of network virtualization technologies. One of the earliest technologies is known by the name *Virtual Local Area Network* (*VLAN*). VLANs are part of the Ethernet standard, and the Ethernet switches used in data centers support the use of VLANs.

A traditional switch without VLANs forms a single network. A set of computers connect to ports on the switch, and all the computers can communicate. When a network administrator uses a VLAN switch, the administrator assigns each port on the switch a small integer known as the port's *VLAN tag*. If a port has VLAN tag X, we say that the computer connected to the port is "on VLAN X." VLAN technology makes the switch hardware behave like a set of smaller switches. Without VLAN technology, any time a computer broadcasts a packet, all computers attached to the switch receive a copy. With VLAN technology enabled, when a computer on VLAN X broadcasts a packet, only computers on VLAN X receive a copy.

An administrator can configure multiple switches that all use VLAN tags consistently. When a packet arrives, hardware adds the port's VLAN tag to the packet, and the tag remains in place when the packet is sent to another switch.

It may seem that VLAN technology suffices for a data center. Indeed, smaller data centers have used VLANs. However, two limits prevent the technology from handling large cloud data centers. First, each VLAN tag consists of twelve bits, which limits the technology to 4096 VLANs. Over four thousand VLANs is sufficient for a single organization because each department can be assigned a separate VLAN number. In a large data center, however, if each tenant is assigned a dozen VLAN tags, the data center is limited to 341 possible tenants. Furthermore, the VLAN scheme assigns VLAN tags to ports on switches, not to VMs or containers. Finally, when they use VLANs, switches run a *Spanning Tree Protocol* that does not scale to large data centers. To summarize:

> *VLAN technology imposes a set of virtual overlay networks on a set of switches, and each computer attached to the switches is assigned to one of the virtual networks. The technology does not solve the problem of separate virtual networks for tenants in a large data center.*

7.5 Scaling VLANs To A Data Center With VXLAN

Network switches are designed to send arbitrary packets among computers, including Internet packets and other types. The ability to communicate over the Internet has become so important that data centers now use Internet packets almost exclusively. The addresses used on Internet packets are known as *IP addresses* (*Internet Protocol Addresses*). Switches throughout each data center are configured to examine IP addresses and forward packets to their correct destination, either within the data center or on the global Internet.

Engineers observed that Internet packets could be used to extend VLAN technology to a scale sufficient for the largest data centers. They devised a technology known as *Virtual Extensible LAN* (*VXLAN*) that many data centers use. The technology requires each switch to include VXLAN software, and requires the network administrator to configure special routing protocols so the VXLAN system can learn the locations of computers in the data center. Once it has been configured, VXLAN can provide the equivalent of more than sixteen million virtual networks —enough for even the largest cloud providers.

VXLAN uses multicast technology to make delivery of packets efficient. To understand how it works, imagine a dozen VMs all on the same VLAN. Suppose three VMs are running on servers on one side of the data center and nine are running on servers on the other side of the data center. If one of the three broadcasts a packet, multicast sends copies to each of them locally. Instead of sending nine individual copies across the data center to each of the other nine VMs, however, VXLAN sends a single copy across the data center which is then distributed to the nine recipients. The use of multicast reduces the amount of traffic VXLAN imposes on the data center network, thereby allowing the technology to scale.

The VXLAN technology reverses the usual network paradigm. A conventional network uses Ethernet packets to carry Internet packets. That is, each Internet packet is placed inside an Ethernet packet for transfer from a computer to a switch or from one switch to another. We use the term *encapsulation* to refer to the idea. VXLAN adds an interesting twist: once an Ethernet packet has been created, VXLAN places the entire Ethernet packet inside an Internet packet for transfer. The surprise is that to send the resulting Internet packet, VXLAN places it in an "outer" Ethernet packet. The extra encapsulation may seem unnecessary, but the effect is that VXLAN can scale VLAN technology to large data centers. To summarize:

> *VXLAN technology uses Internet packets to scale the VLAN approach to a size that suffices for a large data center.*

7.6 A Virtual Network Switch Within A Server

The use of VMs and containers complicates data center networking in two ways. First, unlike a traditional network that assigns an IP address to each physical computer, a data center usually assigns a separate IP address to each virtual machine. Thus, when multiple VMs run on the same physical server, multiple addresses will be assigned to the server. Second, if two VMs in the same server communicate, packets must be forwarded from one to the other (provided the forwarding rules permit such communication).

How can a hypervisor forward packets among the VMs it has created? One answer lies in a *virtual network switch*, such as *Open vSwitch*. The idea is straightforward: use a piece of software that acts like a conventional network switch. Use the physical network connection on the server to connect the virtual switch to the data center network. Connect each VM to the virtual switch, and allow the VM to send and receive packets, just as if the VM connected to a real network switch. More important, arrange for the virtual switch to use standard configuration software, and configure the virtual switch to follow the same rules for packet forwarding as other data center switches (e.g., prevent packets from a VM owned by one tenant from flowing to a VM owned by another). To summarize:

> *A piece of software known as a* virtual network switch *allows a server to assign each VM its own IP address and forward packets among the VMs and the data center network; a virtual switch can be configured to follow the same forwarding rules as other data center switches.*

7.7 Network Address Translation (NAT)

Recall that container technology allows an app to run in a separate environment, isolated and protected from other containers. The question arises, does each container have its own IP address, or do containers all share the IP address of the host OS? The answer is that container technology supports three possibilities:

- A container can clone the host's IP address
- A container can receive a new IP address
- A container can use address translation

A container can clone the host's IP address. If a container uses the same IP address as the host OS, we say that the container has *cloned* the address. Using the same address as the host OS means the container's network use may conflict with the use by normal apps or other containers that have cloned the address. For example, if two containers clone the host's IP address and both attempt to run a web server on port 80, only the first to request the port will succeed. Consequently, cloning is seldom used in a cloud data center.

A container can receive a new IP address. Each container can be assigned a unique IP address, and the host operating system can use a virtual switch to provide connectivity, as described in the previous section.

A container can use address translation. Address translation was invented and widely deployed before cloud computing arose. In fact, many individuals have used NAT technology either from a *wireless router* in their residence or a Wi-Fi hot spot, such as those in coffee shops and hotels. Known by the acronym *NAT* (*Network Address Translation*), the technology allows multiple computers to share a single IP address.

When used with containers, NAT software runs in the host operating system. When a container that uses NAT begins execution, the container requests an IP address, and the NAT software responds to the request by assigning an IP address from a set of reserved, *private* IP addresses that cannot be used on the Internet. How can a container with a private address communicate with computers on the Internet? The NAT software handles communication by intercepting each outgoing packet and replacing the private IP address with the IP address of the host OS. NAT then sends the packet. When a reply comes back, NAT replaces the host's address with the private address assigned to the container, and forwards the reply to the container. Although our description omits several technical details, it describes the effect: NAT allows multiple containers to share the host's IP address safely analogous to the way a wireless router or Wi-Fi hot spot allows multiple devices to share an IP address.

> *A container can clone the host's IP address, be assigned a unique IP address, or use Network Address Translation (NAT) technology to operate safely with other containers running on the same server.*

7.8 Managing Virtualization And Mobility

Managing a conventional network is relatively straightforward because the network and the devices connected to the network remain relatively stable. Of course, the owner must configure each switch by specifying a set of rules the switch will use to process packets. Once a configuration has been entered, however, the network will perform correctly until a piece of equipment fails. Routing software can handle some failures automatically by changing the rules in switches to send packets along alternative paths that avoid the equipment that has failed.

Configuring and managing the network in a cloud data center poses a complex challenge for three reasons:

- Complex interconnections among switches

- Multiple levels of virtualization

- Arbitrary placement of addressable entities and migration

Complex interconnections among switches. To understand the complexity of a physical network, look again at Figure 4.7†. Imagine configuring each switch to have correct forwarding rules for destinations on the global Internet as well as for each possible destination in the data center. Consider that the configuration in a switch must specify all equal-cost paths for a destination, allowing ECMP hardware to balance the load across all the paths.

Multiple levels of virtualization. Consider the configuration needed for VXLAN technology. Each switch in the data center must be configured with an IP address, and IP forwarding must be working correctly before VXLAN can be added. Furthermore, IP multicast must also be configured, and operating correctly because VXLAN uses IP multicast. In addition, VXLAN requires a manager to configure routing protocols that propagate address information across the entire data center.

Arbitrary placement of addressable entities and migration. A provider can place addressable entities —VMs and containers —on arbitrary physical servers. As a consequence, the IP addresses that belong to a given tenant may be spread across the data center. More important, whatever system is used to connect all the virtual entities owned by a tenant must accommodate VM migration, and must update forwarding over the network to allow them to reach one another after a migration. The point is:

> *The complex interconnections, multiple levels of virtualization, arbitrary placement of addressable entities, and VM migration make configuring and operating a data center network especially difficult.*

7.9 Automated Network Configuration And Operation

How can thousands of switches arranged in a complex interconnection be configured correctly and efficiently? How can the forwarding information be updated when a VM moves? The answers lie in technologies that automate the configuration and operation of networks. Examples include:

- Spanning Tree Protocol

- Standard routing protocols

Spanning Tree Protocol. The Ethernet network technology used in data centers allows a sender to broadcast a packet. When a set of switches are connected in a cycle, broadcast causes a potential problem because one switch forwards a copy to another,

---

†Figure 4.7 can be found on page 49.

which forwards to another, and eventually the original switch receives a copy, causing the cycle to repeat forever. Unfortunately, the leaf-spine network architecture used in data centers contains many cycles. The *Spanning Tree Protocol* (*STP*), which runs when a switch boots, solves the problem by detecting cycles and setting up rules to prevent packets from cycling forever.

Standard routing protocols. Like most networks, data center networks employ standard routing protocols that propagate routing information automatically. The protocols, including *OSPF* (*Open Shortest Path First*) and *BGP* (*Border Gateway Protocol*), learn about possible destinations inside the data center and on the global Internet, compute a shortest path to each destination, and install forwarding rules in switches to send packets along the shortest paths. More important, routing protocols monitor networks, and when they detect that a switch or link has failed, automatically change forwarding to route packets around the failure.

7.10 Software Defined Networking

A technology for automated network mangement known as *Software Defined Networking* (*SDN*) stands out as especially important for data centers. SDN allows a manager to specify high-level policies†, and uses a computer program to configure and monitor network switches according to the policies. That is, instead of relying on humans, a data center owner can use software to master the complexity and handle the necessary low-level details. Software can accommodate a large data center, can handle multiple levels of virtualization, and can update forwarding rules when VMs migrate.

The SDN approach uses a dedicated computer to run SDN software. The computer runs a conventional operating system, typically Linux, an *SDN controller* app, and a management app (or apps). A management app can use policies to choose how to forward packets. The controller forms a logical connection to a set of switches and communicates appropriate forwarding rules to each switch. Figure 7.2 illustrates an SDN controller communicating with six switches.

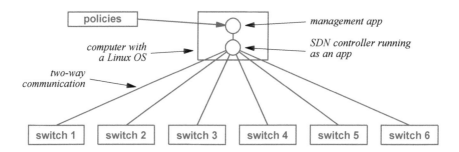

Figure 7.2 An illustration of the communication between a management app, an SDN controller, and a set of network switches being managed.

---

†We say a network manager expresses a set of *intents* and the SDN software translates the intents into configurations that achieve the intended effect.

As the figure shows, the logical connections between an SDN controller and each switch employ bidirectional communication that allows data to flow in either direction. In addition to the controller sending configuration to the switch, the controller can monitor the status of the switch itself and the links to other switches. For example, SDN can configure a switch to inform the controller when events occur, such as link failure or recovery. Furthermore, a switch can send specified packets to the controller. Typically, a controller configures each switch to send the controller any packet that cannot be handled by its forwarding rules. Thus, the controller will be informed of exceptions, which can include packets from a new application or packets sent as an attack on the data center.

7.11 The OpenFlow Protocol

A key component of SDN technology, the *OpenFlow* protocol standard defines the communication available to an SDN controller. That is, OpenFlow specifies both the form and meaning of messages that can pass between a controller and a switch. To use SDN, a switch needs an OpenFlow module; the switches used in data centers include such support.

How does OpenFlow control packet forwarding? The basic idea is straightforward: a controller installs a set of forwarding rules in each switch. Each rule describes a particular type of packet and specifies the output port over which the packet should be sent. When a packet arrives at the switch, the switch hardware checks each of the rules, finds one that matches the packet, and sends the packet over the specified port.

A forwarding rule uses items in a packet header to decide where the packet should be sent. For example, a rule can examine fields that specify the application being used, directing packets that carry World Wide Web traffic out one port, packets that carry database traffic out another port, and all other packets out a third port. Alternatively a forwarding rule can examine the packet's destination, directing packets destined for the global Internet out one port and packets destined for servers inside the data center out another port. A forwarding rule can use the sender's address to direct traffic from some senders along one path and traffic from other senders along another path (e.g., to give some customers priority). Figure 7.3 lists a few of the header fields that a switch can examine when making a forwarding decision.

Field	Meaning
VLAN tag	The VLAN to which the packet belongs
IP source	The sending computer
IP destination	The ultimate destination
TCP source	The type of the sending application
TCP destination	The type of the receiving application

Figure 7.3 Examples of header fields that a controller can use in a forwarding rule.

7.12 Programmable Networks

As described above, the first generation of SDN software uses static forwarding rules, where each rule specifies a set of header fields in a packet and an output port for packets that match the rule. The chief limitation of static forwarding rules lies in the overhead needed to handle exceptions: a packet that does not match any rule must be sent to the controller†. When it receives an exception, the controller examines the policies, chooses an action to use (which may include dropping the packet), installs new forwarding rules in the switches, and then sends the packet back to the switch for processing according to the new rules.

A second generation of SDN has been designed to reduce the need for a controller to handle each exception by placing a computer program in each switch; we use the term *programmable network* to describe the idea. Instead of a conventional language, the second generation uses a special programming language named *P4*. The point is:

> *The second generation of SDN technology allows an SDN controller to install a computer program written in the* P4 *language in each switch. The use of a program means a switch can handle exceptions locally without sending each exception to the controller.*

7.13 Summary

Data center networks use virtualization technologies to give each tenant the illusion of having an isolated network. We say that a virtual network forms an overlay and the physical network on which it has been built forms an underlay. Having multiple layers of overlays on top of other overlays makes configuring and operating a data center network especially complex.

Several virtualization technologies have been used in data centers. Smaller data centers can use VLANs to isolate tenants. For larger data centers with many tenants, an extended VLANs technology named VXLAN must be used; VXLAN uses Internet packets, which means it builds an overlay on top of the Internet protocol, which builds an overlay on top of the physical network. Virtual switch technology and NAT are used to handle forwarding among addressable entities (VMs and containers) that run on a given server.

To manage complex interconnections among data center switches, multiple layers of virtualization, and the placement of addressable entities on arbitrary servers, data center networks use automated configuration and management technologies, including the Spanning Tree Protocol (STP) and standard routing protocols. Software Defined Networking (SDN) allows a manager to run a computer program called a controller, that configures and monitors network switches. A second generation of SDN uses the P4 language to create a programmable network.

---

†Industry uses the term *slow path processing* to refer to packets sent to the controller.

Chapter Contents

8

Virtual Storage

8.1 Introduction

Chapter 4 describes the infrastructure used in data centers, including the idea of separating disks from servers. Other chapters in this part of the text describe virtualization technologies that run on top of the infrastructure, including virtual machines, containers, and virtual networks.

This chapter completes the description of virtualization technologies by examining the virtual storage facilities used in data centers. The chapter introduces the concepts of remote block storage, remote file storage, and the facilities used to provide them. The chapter also discusses object storage (also known as a key-value storage).

8.2 Persistent Storage: Disks And Files

We will learn that the storage facilities used in data centers employ the same designs as the storage mechanisms used on a conventional computer. In fact, data center storage mechanisms reuse approaches that have been around for decades. Therefore, to understand persistent storage in a data center, we must start with the persistent storage systems that are used with conventional computer systems.

The term *persistent storage* (or *non-volatile storage*) refers to a data storage mechanism that retains data after the power has been removed. We can distinguish between two forms of persistent storage;

- Persistent storage devices
- Persistent storage abstractions

Persistent storage devices. A conventional computer uses a separate physical device to provide persistent storage. By the 1960s, the computer industry adopted electromechanical devices called *disks* that use magnetized particles on a surface to store data. The industry now uses Solid State Disk (*SSD*) technology with no moving parts.

Persistent storage abstractions. Users do not deal directly with disk hardware. Instead, an operating system provides two abstractions that users find intuitive and convenient: *named files* and *hierarchical directories*, which are also known as *folders*. The file abstraction offers two important properties: files can vary in size and can store arbitrary data (e.g., a short text document, a spreadsheet, a photo, or a two-hour movie). The hierarchical directory abstraction allows a user to name each file and to organize files into meaningful groups.

8.3 The Disk Interface Abstraction

A disk device provides a *block-oriented interface*. That is, the hardware can only store and retrieve fixed-sized blocks of data. Traditional disks define a block to consist of 512 bytes of data; to increase performance, some newer disks offer blocks of 4096 bytes†. The blocks on a disk are numbered starting at zero (0, 1, 2,...). To store data on a disk, the operating system must pass two items to the disk device: a block of data and a block number. The disk hardware uses the block number to find the location on the disk, and replaces the contents with the new values. We say the hardware *writes* the data to the specified block on disk. The hardware cannot write a partial block —when the operating system instructs the disk device to store data in the ith block, the entire contents of the ith block on the disk is replaced with the new data. Similarly, when an operating system instructs a disk device to retrieve a copy of a previously-stored block, the operating system must specify a block number. The disk hardware always fetches a copy of an entire block. Figure 8.1 summarizes the operations that the hardware provides.

Operation	Parameters	Meaning
read	block number	Fetch a copy of specified block from the disk
write	block number and block of data	Store the specified block on disk, replacing the previous value

Figure 8.1 The operations in the block-oriented interface that a disk provides. The hardware always transfers a complete block.

The disk interface is surprisingly constrained: the hardware can only transfer a complete block, and does not transfer more than one block per request. A later section explains how data centers use a block-oriented interface.

---

†The storage industry uses powers of 2 for block sizes.

8.4 The File Interface Abstraction

An operating system contains a software module known as a *file system* that users and applications use to create and manipulate files. Unlike a disk interface, a file system provides a large set of operations that the operating system maps onto the underlying disk hardware. Unfortunately, each operating system defines many details, such as the format of file names, file ownership, protections, and the exact set of file operations. Nevertheless, most systems follow an approach known as *open-close-read-write*. Figure 8.2 lists the basic operations the interface provides.

Operation	Meaning
open	Obtain access to use a file and move to the first byte
close	Stop using a previously-opened file
read	Fetch data from the current position of an open file
write	Store data at the current position of an open file
seek	Move to a new position in an open file

Figure 8.2 File operations that an *open-close-read-write* interface provides.

In practice, a given file system will offer additional operations, such as the ability to change the name of a file, its location in the directory system, file ownership and protections, and a way to obtain metadata, such as the time the file was created and last modified. We will learn that differences among file systems complicate the design of file storage systems in data centers.

The chief difference between the operations used with files and those used with disk hardware arises from the transfer size. Unlike a disk device, a file system provides a *byte-oriented interface*. That is, the interface allows an application to move to an arbitrary byte position in a file and transfer an arbitrary number of bytes starting at that position. We can summarize:

> *The interfaces used for persistent storage mechanisms follow one of two abstractions. Disk hardware uses a block-oriented abstraction, and file systems use a byte-oriented abstraction.*

8.5 Local And Remote Storage

We use the term *local storage device* to characterize a disk connected directly to a computer. Industry also uses the term *Directly Attached Storage* (*DAS*). The connection occurs over a piece of hardware known as an *I/O bus*, and all interaction between the processor and the disk occurs over the bus. Both electromechanical disks and solid state disks connect over a bus, and either can provide local storage.

We use the term *remote storage* to characterize a persistent storage mechanism that is not attached directly to a computer, but is instead reachable over a computer network†. A disk device cannot connect directly to a network. Instead, the remote disk connects to a *storage server* that connects to a network and runs software that handles network communication. Figure 8.3 illustrates the architecture.

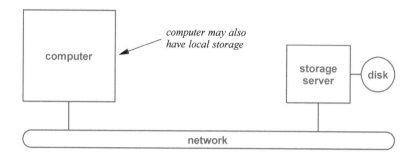

Figure 8.3 Conceptual organization of a *remote storage* facility with a computer accessing a storage server over a network and disk storage attached to the storage server.

8.6 Two Types Of Remote Storage Systems

The first remote systems appeared in the 1980s, shortly after Local Area Networks (LANs) became available and almost two decades before cloud data centers were invented. Interestingly, commercial systems for each of the two remote storage paradigms appeared within the span of a few years:

- Byte-oriented remote file access
- Block-oriented remote disk access

Byte-oriented remote file access. In the 1980s, as organizations moved from shared departmental computers to individual workstations that each had local storage, sharing files became inconvenient. When a group collaborated on a document, they had to send copies to each other. To solve the problem, computer vendors introduced remote file storage. The idea is straightforward: equip a storage server with a disk, and arrange for the storage server to run an operating system that includes a file system. Add software to each individual's workstation that can access and modify files on the server. That is, an app running on the user's computer can use the open-close-read-write paradigm to manipulate a remote file. Each time an app performs an operation on a remote file, the user's operating system sends the request to the server, which performs the operation on the file.

One of the first remote file access systems is known as the *Network File System* (*NFS*). Originally devised by Sun Microsystems, Inc., NFS became an Internet stan-

---

†To achieve high performance, the network must have low latency and high throughput.

dard, and is still used, including in data centers. NFS integrates the remote files into the user's local file system. That is, some folders on the user's computer correspond to a remote file system. We say that NFS provides *transparent* remote file access because a user cannot tell the location of a file, and an app can access a remote file as easily as a local file.

Block-oriented remote disk access. In the 1980s, systems known as *diskless workstations* became commercially available. Such systems do not have local storage; they require the owner to purchase a storage server that manages a disk for the workstation to use. When it needs to fetch or store a disk block, the operating system on the diskless workstation sends a request over the network to a server. The server performs the requested operations on behalf of the diskless workstation and sends a response. When it sends a *write* request, the diskless workstation specifies both a block number and the data to be written to the block; the reply from the server either announces success or contains an error code to tell why the operation could not be performed. When it sends a *read* request, the diskless workstation only specifies a block number; the reply either contains a copy of the requested block or an error code to tell why the operation could not be performed.

We can summarize:

> *Commercial remote storage systems first appeared in the 1980s. Both remote file access mechanisms and remote disk access mechanisms were in wide use before cloud data centers appeared.*

8.7 Network Attached Storage (NAS) Technology

Three implementations of storage servers have been used:

- Host-based
- Server-based
- Specialized hardware

Host-based. A storage server consists of a computer that has network access, directly-attached local storage, and software to handle requests that arrive over the network. In principle, any computer can act as a storage server once server software has been loaded. Most operating systems already include software that allows a user to share files. We use the term *host-based* to characterize a conventional host computer that runs file sharing software. A host-based server has the advantage of not needing expensive hardware, but the disadvantages of low performance and limited scale.

Server-based. Using dedicated, high-speed server hardware can increase performance and allow a storage server to scale beyond the capabilities of a host-based system. Increased processing power (i.e., more cores) allows a server to handle many requests per second and a large memory allows the server to cache data in memory. In addition, a server with a high-speed bus can accommodate more I/O, including multiple local disks.

Specialized hardware. To be useful in a cloud data center, a file server must offer both high performance and durability. The performance requirements are obvious: a data center has many tenants, and each tenant may choose to share files among many VMs and containers. The durability requirement arises because storage servers used in a data center operate continuously and downtime can be costly. Therefore, cloud providers look for systems where the *Mean Time To Failure* is long.

Industry has adopted the term *Network Attached Storage (NAS)* to refer to specialized systems that provide scalable remote file storage systems suitable for a data center. The hardware used in a NAS system is ruggedized to withstand heavy use. The hardware and software in a NAS system are both optimized for high performance.

One technique used to help satisfy the goal of durability involves the use of parallel, redundant disks. Known as a *RAID* array (*Redundant Array of Independent Disks†*), the technology places redundant copies of data on multiple disks, allowing the system to continue to operate correctly if a disk fails, and allows the disk to be replaced while the array continues to operate (known as *hot swapping*). To summarize:

> *Network Attached Storage (NAS) systems use specialized hardware and software to provide remote file storage. The high performance and durability of NAS systems make them suitable for use in data centers.*

8.8 Storage Area Network (SAN) Technology

Industry uses the term *Storage Area Network (SAN‡)* to describe a remote storage system that employs a block-oriented interface to provide a remote disk interface. Analogous to NAS, the term SAN implies that the system has optimized, ruggedized hardware and software that provide the performance and durability needed in a data center.

Early SAN technology includes two components: a server and a network optimized for storage traffic. Some of the motivation for a special network arose because early data centers used a hierarchical network architecture optimized for web traffic (i.e., north-south traffic). In contrast to web traffic, communication between servers and remote storage facilities produces east-west traffic. Thus, having a dedicated network used to access remote storage keeps storage traffic off the main data center network.

A second motivation for a separate storage network arose from capacity and latency requirements. In terms of capacity, a hierarchical data center network means increasing the capacity of links at higher levels. As Chapter 4 explains, network technologies are only available for a fixed set of capacities, so it may not be possible to purchase the additional capacity needed for storage traffic. In terms of latency, a hierarchical data center network means storage traffic must flow up the hierarchy and back down.

---

†The acronym originally expanded to *Redundant Array of Inexpensive Disks.*

‡Attempting to be clever, the storage community chose the acronyms NAS and SAN to be a pair of semordnilaps (each is the other backwards).

We can summarize:

> *Storage Area Network (SAN) technology provides remote disk access. To handle scale, SANs use specialized equipment. Early SANs use a specialized, dedicated network with low latency and high capacity to keep storage traffic separate from other traffic.*

8.9 Mapping Virtual Disks To Physical Disks

How does a SAN server work? The server has one or more local disks that it uses to store blocks on behalf of clients. The server does not merely allocate one physical disk to each client. Instead, the server provides each client with a *virtual disk*. When software creates an entity that needs disk storage (e.g., when a VM is created), the software sends a message to the SAN server giving a unique ID for the new entity and specifying a disk size measured in blocks.

When it receives a request to create a disk for a new entity, the server uses the client's unique ID to form a *virtual disk map*. The map has an entry for each block of the virtual disk, 0 1, and so on. For each entry, the server finds an unused block on one of the local disks, allocates the block to the new entity, and fills in the entry with information about how to access the block. Basically, the virtual disk map defines how to treat a set of blocks on disks at the server as a single disk. Note that the blocks allocated to an entity may be on one physical disk at the server or spread across multiple physical disks. Figure 8.4 illustrates a few entries in an example virtual disk map.

Client's Block Number	Server's Physical Disk	Block On The Disk
0	27	21043
1	83	8833
2	91	77046
3	15	90023
4	7	16082
5	79	9274
6	32	61084

Figure 8.4 An example of entries in a virtual disk block map†.

To an entity using the SAN server, the server provides the abstraction of a single physical disk. Therefore, the entity will refer to blocks as, 0, 1, 2, and so on. When a *read* or *write* request arrives at the SAN server, the server uses the client's ID to find the map for the client. The server can then use the map to transform the client's block number into the number of a local physical disk and a block on that disk. The server performs the requested operation on the specified block on the specified local disk. The

---

†In practice, the use of RAID technology by a storage server makes the mapping more complex than our simplistic example.

client remains completely unaware that sending a request for block 0 causes the server to access block 21043 on disk 27 and sending a request for block 1 causes the server to access block 8833 on disk 83.

8.10 Hyper-Converged Infrastructure

The specialized networks used in early SANs were expensive. The move to leaf-spine networks and the availability of much less expensive high-capacity Ethernet hardware changes the economics of SANs. Instead of using a special-purpose network, SAN hardware has been redesigned to allow it to communicate over a conventional data center network.

Industry uses the term *converged network* to describe a network that carries multiple types of traffic. To characterize a data center network that carries all types of traffic, including SAN storage traffic, industry uses the term *Hyper-Converged Infrastructure* (*HCI*).

> *Because data centers shifted to the leaf-spine network architecture that has the capacity necessary for storage traffic and the cost of Ethernet hardware fell, SAN producers changed from the use of special-purpose networks to a Hyper-Converged Infrastructure that sends all traffic over the data center network.*

8.11 A Comparison Of NAS and SAN Technology

The question arises: is NAS or SAN better for a cloud data center? The short answer is that neither is always better. Each has advantages in some situations. The following sections highlight some of the advantages and disadvantages of each technology.

8.11.1 NAS Advantages And Disadvantages

A NAS system allows apps to share individual files. A NAS system, such as NSF, can blend remote directories into the local file hierarchy, hiding the differences and making remote file access transparent to a user. Because it presents a normal file system interface with the usual open-read-write-close semantics, NAS works equally well with any app.

One significant advantage for NAS arises from the use of containers. When constructing an image for a container, a programmer can specify external mount points in the container's file system. When a user creates a container to run the image, the user can choose to connect each external mount point to a directory in the host's local file system or to a remote directory on a NAS server.

A final advantage of NAS arises because the actual file system resides on a NAS server. Therefore, only file data passes across the network. The server maintains all metadata (e.g., directories and inodes).

In terms of disadvantages, because each operating system defines its own type of file system, each NAS is inherently linked to a particular OS. Thus, a user cannot choose to run an arbitrary operating system unless it supports the NAS file system. Even if all computers accessing NAS and the NAS server run the same operating system, they must all agree on accounting details. For example, because each file has an owner and access permissions, all computers that access a file must agree on user IDs.

8.11.2 SAN Advantages And Disadvantages

The block access paradigm has the advantage of working with any operating system. A user can create a VM running Windows and another VM running Linux. Using integers to identify blocks has the advantage of making the mapping from a client's block number to a block on a local disk extremely efficient.

Of course, the remote disk paradigm has some disadvantages. Because each entity has its own virtual disk, entities cannot share individual files easily. Furthermore, although it works well for virtual machines, containers cannot use a block-oriented interface directly. Finally, because a file system resides on an entity using a SAN, the file system must transfer metadata over the network to the SAN site.

8.12 Object Storage

The previous sections describe two types of cloud storage technologies: block storage that uses the remote disk paradigm (SAN) and byte-oriented storage that uses the remote file paradigm (NAS). The two paradigms seem to suffice for all needs. A remote disk system allows an operating system to run the same as if the computer has a local disk attached. A remote file system can be used to store and share any digital object, including documents, images, recordings, and web pages. However, a remote file system mechanism inherently depends on a particular operating system, and can only be accessed by apps that can run on the operating system.

As cloud systems emerged, a question arose: can an alternative storage mechanism be devised that allows arbitrary apps running on arbitrary operating systems to store and share arbitrary data items? The question has been answered by *object store* or *key-value store* technologies†. An object store system has three characteristics that make it especially popular in a cloud environment:

- Stores arbitrary objects
- Offers universal accessibility and scale
- Remains independent of the operating system

Stores arbitrary objects. Like a file system, an object store can contain arbitrary objects, and objects can be grouped together into *buckets*, analogous to file folders.

---

†As an example, Amazon provides a well-known object store known as *Simple Storage Service (S3)*.

Offers universal accessibility and scale. An object store uses a general-purpose interface that any app can use, and scales to allow many simultaneous accesses, including apps in containers. Typically, an object store offers a *RESTful interface.* That is, an object store employs the *http* protocol that a browser uses to access a web page.

Remains independent of the operating system. Unlike a NAS system, an object store does not depend on a specific operating system or file system. Instead, an app running on an arbitrary operating system can access the object store to save and retrieve objects, making it useful for both VMs and containers.

8.13 Summary

A conventional computer system uses two forms of persistent storage: a hardware device (disk) that stores data and a software system that uses the hardware to store files and directories. Disk hardware uses a block interface that allows a computer to read or write one block at a time; a file system typically uses the open-close-read-write paradigm to provide a byte-oriented interface.

Remote storage systems were commercially available for both the block paradigm and file paradigm decades before cloud computing was invented. Each type requires a server that has a network connection and local storage attached. To use a remote disk server, a computer sends requests to read or write a specified block. To use a remote file server, a computer sends requests to read or write a set of bytes to a specified file.

Industry uses the term Network Attached Storage (NAS) to describe a high-performance, scalable, ruggedized remote file server, and the term Storage Area Network (SAN) to describe a high-performance, scalable, ruggedized remote disk server that provides block storage.

Block storage systems allocate a virtual disk to each client (e.g., each VM). The client uses block numbers 0 through N−1. The SAN server maintains a mapping between the block numbers a client uses and blocks on physical disks; the client remains unaware that blocks from its disk may not map onto a single physical disk at the server.

Early SAN technology used a special-purpose, dedicated network to connect computers to remote storage servers. The shift to leaf-spine networks that support high volumes of east-west traffic has enabled a Hyper-Converged Infrastructure (HCI) in which communication between VMs and a storage server uses the data center network along with other traffic.

SAN technology and NAS technology each have advantages and disadvantages; neither satisfies all needs. A SAN can allow a VM to function as if it has a local disk attached. Containers can access files on a NAS server easily, but NAS requires the client to have the same file systems as the server.

Object store technology provides an alternative to NAS and SAN systems. Like a remote file system, an object store can store an arbitrary digital object. Unlike a remote file system, object store technology uses a RESTful interface to provide universal accessibility independent of any file system or operating system.

Part III

Automation And Orchestration

Software Systems That Automate Deployment, And Operation Of Facilities And Services

Chapter Contents

9

Automation

9.1 Introduction

Previous parts of the text discuss the motivation for cloud computing, the physical infrastructure, and key virtualizations that cloud systems use. This part of the book introduces the topics of automation and orchestration, and this chapter explains why cloud systems need automated support mechanisms. It examines aspects of automation, including procedures that have been automated as well as conceptual levels of automation. Most important, it explains why so many automation systems have arisen. The next chapter examines one example of orchestration technology in detail.

9.2 Groups That Use Automation

Automation focuses on making tasks easier and faster for humans to perform. Cloud automation mechanisms have been developed for three categories of users:

- Individual customers
- Large cloud customers
- Cloud providers

Individual customers. Individual subscribers often use SaaS apps, such as a document editing system that allows a set of users to share and edit documents cooperatively. To make such services convenient, providers typically offer access through a web browser or a dedicated app that the user downloads. The interfaces presented to individuals hide details, allowing a user to focus on using the services.

Providers often offer ways for individuals to try more advanced services. For example, to encourage individuals to create web sites, providers offer free trials with step-by-step instructions and a point-and-click web interface. When a user makes a request, the provider may create a virtual machine or container, allocate storage, configure network access, and launch web server software. The point is that when an individual uses a point-and-click interface to access a service, the interface must be backed by underlying automated systems that handle many chores on behalf of the individual.

Large cloud customers. Unlike a typical individual, an enterprise company or other organization that moves to a public cloud needs tools to control and manage computing. Two types of automated tools are available for large cloud customers. One type, available from the provider or a third party, allows a customer to download and run the tools to deploy and manage apps. The next chapters explain examples, including Kubernetes, which automates deployment and operation of a service built with containers, and Hadoop, that automates MapReduce computations. The other type consists of tools offered by a provider that allow large customers to configure, deploy, and manage apps and services without downloading software.

Cloud providers. Cloud providers have devised some of the most sophisticated and advanced automation tools, and use them to manage cloud data centers. The next sections consider the scope of the problems that providers face and the ways automation can help. When thinking about automated tools, remember that in addition to building tools to configure, monitor, and manage the underlying cloud infrastructure, a provider also creates tools that handle requests from cloud customers automatically. Tools are available that accommodate requests from both individual customers and large organizational customers.

9.3 The Need For Automation In A Data Center

Consider the tasks of building and operating a large data center. In each case, the effort can be daunting. After installing a raised floor, air conditioning facilities, and thousands of racks, equipment must be loaded into each rack. Each piece of equipment must be connected to both a power source and to the data center network. However, such installation is only the beginning. Each piece of equipment must be assigned a network address, and the switches must be configured to forward packets along a shortest path from any source to any destination, including destinations outside the data center. Each server must be configured to run appropriate software. For example, a server that will host VMs must be configured to run a type 1 hypervisor.

After all the facilities have been installed and configured, operating a data center is much more complex than operating IT facilities for a single organization. Four aspects of a cloud data center stand out.

- Extreme scale
- Diverse services
- Constant change
- Human error

Extreme scale. A cloud provider must accommodate thousands of tenants. Although some tenants will be individuals, others will be major enterprise customers. Each enterprise customer may deploy dozens of VMs and hundreds of containerized services. The total number of services can become extremely large.

Diverse services. Unlike an enterprise, a cloud data center provider allows each customer to choose software and services to run. Consequently, the cloud data center may run software for thousands of services at the same time.

Constant change. A data center provider must handle dynamically changing requirements, with the response time dependent on the change. At one extreme, when a new tenant appears, the tenant must sign a contract and plan how to migrate its data and computation to the cloud, which gives the provider ample time to configure facilities for the new tenant. At the other extreme, a tenant can request a new VM or deploy a container at any time, and the data provider must accommodate the request quickly.

Human error. Data center operators report that:

Many problems in a data center can be traced to human error.

9.4 An Example Deployment

To understand data center operations, consider a simplistic example: deploying a VM. Previous chapters describe some of the details. Figure 9.1 lists example steps a provider takes when deploying a single VM.

Step	Action Taken
1.	Choose a server on which to run the VM
2.	Configure the hypervisor on the server to run the VM
3.	Assign an IP address to the VM
4.	Configure the network to forward packets for the VM, which may involve configuring the tenant's virtual network (e.g., VXLAN)
5.	Choose a remote disk server and allocate storage for the VM, according to the tenant's specification
6.	Configure the hypervisor to send requests from the VM to the storage server

Figure 9.1 Example steps a provider may take when deploying a VM.

It should be obvious that performing each step manually will require a significant amount of time. Consequently, automation is needed to perform operations quickly and to accommodate the scale of a data center. We can summarize:

Because even a trivial procedure, such as deploying a VM, requires multiple configuration steps, manual operation does not suffice — operating an entire data center efficiently requires automation.

9.5 What Can Be Automated?

The short answer is that because they involve the management of data and computer software, almost all operational tasks in a data center can be automated. Recall, for example, that some data centers use a "lights-out" approach in which automated systems run the data center.† The previous chapters describe many facilities and services that a data center provider must manage. The following lists example items that can be automated, and provides a short summary for each.

- **Creation and deployment of new virtual resources**

  The creation of new virtual machines and containers; new virtual storage facilities, including virtual disk images (SAN) and initial contents of virtual file systems (NAS); new virtual networks, including VLANS, IP subnets, IP forwarding, IP multicast, and extended VLANs (VXLAN).

- **Workload monitoring and accounting**

  Measurement of the load on servers, storage facilities, and networks; tracking each tenant's resource use and computing charges; identification of hot spots; long-term trends, including capacity assessment and predictions of when additional physical facilities will be needed.

- **Optimizations**

  Optimizations for both initial deployments and subsequent changes; the initial placement of VMs and containers to handle balancing the load across physical servers; minimization of the latency between applications and storage, and minimization of network traffic; VM migration, including migration to increase performance or to minimize power consumption.

- **Safety and recovery**

  Scheduled backups of tenant's data; server monitoring; monitoring of network equipment and fast re-routing around failed switches or links; monitoring of storage equipment, including detecting failures of redundant power supplies and redundant disks (RAID systems); automated restart of VMs and containers; auditing and compliance enforcement.

- **Software update and upgrade**

  Keeping apps and operating system images updated to the latest versions; upgrading to new releases and versions of software as specified by a tenant; providing facilities

---

†A description of lights-out operations can be found on page 41.

a tenant can use to update their private software and deploy new versions; aid in achieving continuous deployment of a tenant's apps.

• **Administration of security policies**

Deploying network security across the data center in accordance with the provider's policies, including firewall facilities; protecting each tenant's data and computation; facilities for the management of secrets and encryption keys.

9.6 Levels Of Automation

Various models have been devised to describe the extent to which automation can be used in a data center. Although no single model can capture all the possibilities, a basic model can help explain the extent to which automation can be applied. Figure 9.2 breaks automation into five levels (plus a zero level for completely manual operation).

Level	Description
5	**Automated remediation of problems**
4	**Automated identification of root causes**
3	**Automated analysis of trends and prediction**
2	**Automated monitoring and measurement**
1	**Automated preparation and configuration**
0	**No automation (manual operation)**

Figure 9.2 An example model with five levels of automation. Successively higher numbers denote increased automation.

Level 1: Automated preparation and configuration. Level 1 refers to the automation of tasks that are performed before installation occurs. Industry sometimes uses the term *offline* to refer to such tasks because they can occur before computation begins.

Level 2: Automated monitoring and measurement. Level 2 automation refers to monitoring a data center and making measurements available to human operators. The items that can be monitored include both physical and virtual resources. For example, in addition to servers, network links and switchs, and the disks used in storage facilities, a data center owner usually monitors the power and cooling equipment. Monitoring often focuses on performance, and includes the load on each server, the traffic on each link in the network, and the performance of storage systems.

Level 2 monitoring often includes mechanisms to alert a data center operator when an outage occurs or when the load on a given resource exceeds a specified level. For example, an alert might be raised if a SAN storage system crashes or the load on a network link exceeds 95% of the link capacity.

Level 3: Automated analysis of trends and prediction. Level 3 automation enhances level 2 monitoring by adding analytic capabilities. Instead of merely using threshold values to trigger alerts, a level 3 system can collect measurements over a long period and use software to analyze changes and trends. For example, suppose that the average load on a given network link increases at a rate of 3% each month. As long as the value remains below the alert threshold, a level 2 system will not report a problem. Furthermore, even if a human operator looks at a report of link loads each month, the operator may not spot the trend amid all the other links. Analytic software used at level 3 makes it possible to monitor thousands of links without missing long-term trends. From a data center owner's point of view, the key advantage of level 3 analysis lies in the ability to predict needs, allowing the data center owner to plan ahead rather than waiting for a crisis to occur.

Level 4: Automated identification of root causes. Level 4 automation uses data gathered from monitoring along with knowledge of both the data center infrastructure and layers of virtualization that have been added to deduce the cause of problems. Deduction implies that level 4 automation employs *Artificial Intelligence (AI)* technologies that can understand relationships and reason about cause and effect.

As an example of level 4, consider a situation in which three events occur at the same time: communication between a container and its remote file storage (e.g., a file stored on NAS) suddenly stops working, a tenant's application can no longer access the tenant's database, and a newly created VM cannot access a SAN disk to boot an operating system. A level 4 automation can search for commonalities and outages that can explain all three events. Are the inaccessible systems all located in one part of the data center and, if so, are other devices in the area accessible? Have any network link failures been detected that would account for the three events? Are systems involved on the same extended VLAN? Are the inaccessible systems on the same IP subnet? Has the network been reconfigured or network routing been updated? Has access been terminated for security reasons or because a tenant forgot to pay their bill? Once it identifies a list of the most probable causes, a level 4 system can inform the data center operator, who can make a repair quickly.

Level 5: Automated remediation of problems. Level 5 automation extends the ideas of a level 4 system by adding automated problem solving. Remedial actions may be straightforward (e.g., restarting a VM that has crashed or rebooting a storage server) or complicated (e.g., running a program to reconfigure network forwarding for a particular destination or a set of destinations).

Of course, an automated system may be incapable of solving all problems in a data center. If a widespread power failure occurs, an electrician may be required to diagnose the problem and restore power. Similarly, if a server or network switch fails, an automated system can impose a temporary solution by moving computation to another server or routing network traffic around the failed switch. However, until robots become available that can replace defective equipment, a human is required to provide a permanent solution.

9.7 AIops: Using Machine Learning And Artificial Intelligence

Higher levels of automation require sophisticated software systems. For example, Levels 3 and above may use *machine learning* (*ML*) software. The top levels may use additional forms of *Artificial Intelligence* (*AI*).

Industry uses the term *AIops* (Artificial Intelligence operations) to describe an automation system that uses AI and can operate a data center without human intervention. When AIops was first proposed, the task of operating a data center seemed too complex for AI to handle. However, AI automation systems continue to gain capabilities.

9.8 A Plethora Of Automation Tools

Dozens of tools and technologies have been created to automate various aspects of data center operations. Many questions arise concerning the tools. Which is the best? Does a given data center need to use more than one? Do they all work well together, or are there cases where two or more tools attempt to follow conflicting choices?

Perhaps the most puzzling question is: "Why have so many tools been designed?" The answer lies in two observations:

- Operating a data center is an extremely complex task
- It is easiest to automate each small part of the task independently

Management complexity. As we have seen, data center operations encompass a broad set of facilities and services, both physical and virtual. In addition, a data center owner must manage a broad variety of computation, networking, and storage mechanisms in the presence of continuous change. Despite the breadth of services and large scale, complexity also arises because no optimum choices exist. For example, consider VM placement. An operator may have multiple goals. One goal might focus on avoiding hot spots. To achieve the goal, a new VM should be placed on a lightly-loaded server. A second goal might focus on minimizing network traffic. To achieve the goal, a new VM should be placed near other VMs owned by the same tenant. A third goal might focus on keeping VMs near the storage server the VM will use. A fourth goal might focus on placing active VMs on a subset of servers, making it possible to reduce power costs by powering down some servers. Although each goal seems laudable, finding a location that optimizes all of them simultaneously may be impossible. The point is:

A data center owner may have multiple, conflicting goals. Even with an automated system, it may be impossible to find a way to satisfy all goals simultaneously.

Automating each small task independently. Although it may be impossible to build an automation that optimizes all goals, it is possible to design small tools that each help automate one small task. Such tools can be especially useful if they relieve humans from tasks that involve tedious details. Human error is a source of many problems, and a tool is less prone to making errors. For example, a tool does not make typing errors when configuring a device or service.

> *Many data center automation tools exist because each tool handles one small task. Small tools work well for tasks that involve details because using a tool can reduce errors.*

9.9 Automation Of Manual Data Center Practices

Another reason so many data center automation tools exist arose as a consequence of earlier operational procedures. That is, many tools are designed to automate tasks that humans had been performing manually. To understand the tools, one must know how humans operated data centers. Figure 9.3 illustrates an example workflow for manual configuration.

Step	Action
1	A tenant signs a contract for a new VM and a new work order (i.e., a "ticket") is created
2	A human from the group that handles VMs configures a new VM and passes the ticket on
3	A human from the group that handles networking configures the network and passes the ticket on
4	A human from the group that handles storage configures a SAN server and passes the ticket on
5	The tenant is notified that the VM is ready

Figure 9.3 An example workflow for manual configuration of a VM.

A key principle underlies the division into steps: a given human will not have expertise across all aspects of data center operations. As the example illustrates, steps in the workflow are chosen to allow each step to be handled with limited expertise. Thus, a human who configures the network does not need to know how to configure a VM or storage.

The example workflow helps explain another aspect of automation tools, their limited scope. Tools are created to help humans perform operational tasks. Because each human operator has limited expertise, a tool designed to help an operator focuses on one aspect of data center operations. As a result, although tools help automate each step, the overall workflow remains the same. We can summarize:

Because automation tools evolved to help human operators who each have limited expertise, each tool tends to focus on one aspect of data center operations.

9.10 Zero Touch Provisioning And Infrastructure As Code

One particular use of automation has emerged as necessary for large scale: automated configuration. As we have seen, even a trivial operation, such as deploying a VM, requires a significant amount of configuration. A cloud provider must configure servers, networks, storage, databases, and software systems continuously. Furthermore, each vendor creates their own specialized configuration language, and a data center contains hardware and software from many vendors. Consequently, many automation systems have been devised to handle configuration.

The first step toward automated configuration is straightforward. Instead of requiring a human to learn details, an automation tool can allow humans to specify items in a vendor-independent language. The tool reads the specification and translates the requirements into the vendor-specific configuration language, and passes the appropriate commands to the hardware or software system being configured. In other words, a tool allows a human to specify the values to be used without specifying how to install those values in the underlying systems. The operator does not need to interact with the system being configured.

Industry uses the term *Zero Touch Provisioning* (*ZTP*) and the more awkward term *Infrastructure as Code* (*IaC*) to refer to a process where a data center operator creates a specification and uses software to read the specification and configure underlying systems. Two approaches have been used: *push* and *pull*. The push version follows the traditional pattern of installing a configuration: a tool reads a specification and performs the commands needed to configure the underlying system. The pull version requires an entity to initiate configuration. For example, when a new software system starts, the system can be designed to pull configuration information from a server.

9.11 Declarative, Imperative, And Intent-Based Specifications

The specifications used with automated tools can take many forms, and the question arises, "What form should be used?" Two aspects have become important:

- Declarative vs. imperative
- Intent-based vs. detailed

Declarative vs. imperative. An *imperative* specification states an action to be performed. For example, an imperative specification to assign an IP address to a VM might have the form:

Assign IP address 192.168.1.17 to the VM's Ethernet interface

Although they may seem to work, imperative specifications follow the paradigm of *early binding* by specifying operations for the underlying system and values to be used. The result can be misleading and ambiguous. What happens, for example, if a VM does not have an Ethernet interface, but instead has some other type of network interface? What happens if a VM has multiple Ethernet interfaces?

A *declarative specification* focuses on the result rather than the steps used to achieve it. For example, a declarative specification for IP address assignment might have the form:

Main IP address: 192.168.1.17

Notice that a well-stated declarative specification handles the case where a VM has a non-Ethernet network interface as well as the case there a VM has multiple network interfaces.

Intent-based vs. detailed. Industry uses the term *intent-based* to characterize a specification that allows a human to state the desired outcome without giving details about how to achieve the desired outcome or specific values to be used. For example, an intent-based specification for IP address assignment might state:

Each VM receives a unique IP address on the tenant's IP subnet

without specifying the IP address to be used on each VM. Intent-based specifications offer generality and flexibility. Because they do not dictate steps to be taken, intent-based specifications allow many possible implementations to be used. Because they do not dictate values to be used, intent-based specifications allow configuration tools the freedom to assign any values that produce the desired outcome and thereby encourage innovation. The point is:

> *Using a declarative, intent-based configuration specification can help eliminate ambiguity and increase both generality and flexibility. An intent-based specification gives tools freedom to choose an implementation that produces the desired outcome.*

9.12 The Evolution Of Automation Tools

Because cloud is new, automation tools and technologies continue to evolve. For example, consider container deployment and networking. Kubernetes, described in the next chapter, provides a large set of tools that can be used to manage containerized software deployments. Various additional tools and technologies have been created to control network communication among containers. By default, Kubernetes assigns a unique IP address to each group of containers (called a *pod*). Doing so means network forwarding can be arranged to allow containers in a group to communicate and run microservices, even if the containers run on multiple servers. Docker software takes the approach of using a virtual layer 3 bridge to allow containers to communicate. Other tools are available that can configure an overlay network for containers such that each host has a separate IP subnet and each container on a host has a unique address. Finally, a tool is available to provide secure network communication for containers and microservices. The point is that even fundamentals, such as container networking, continue to evolve.

9.13 Summary

The diversity of services, large scale, and constant change in a cloud data center mandate the use of automated tools for the configuration and operation of hardware and software. Even a trivial operation, such as creating a new VM, requires multiple steps.

Almost all operational tasks can be automated. Examples include the initial creation and deployment of virtual resources, workload monitoring and accounting, optimizations, safety and recovery mechanisms, software update and upgrade, and administration of security policies.

Several models have been proposed to help identify levels of automation. One example divides automation into five levels: automated configuration, automated monitoring, automated analysis of trends, automated identification of root causes, and automated remediation of problems.

Because operating a data center is complex, providers have multiple, conflicting goals. Even with automation, it may be impossible to satisfy all goals simultaneously. Dozens of automation tools and technologies have been created that each handle one small task; many tools are designed to improve a process humans use when operating a data center manually.

The first step toward automated configuration allows a human to specify configuration in a vendor-independent form, and then uses a tool to read the specification and configure the underlying hardware and software systems accordingly. Industry uses the term Infrastructure as Code (IaC) to refer to the process. IaC can be implemented by pushing a configuration to each underlying system when the specification changes or by arranging each underlying system to pull configuration information from a server when the system boots.

An imperative specification follows a paradigm of early binding by specifying the operations and values to be used for configuration. As an alternative, using a declarative specification can help avoid ambiguities. An intent-based specification, which allows a human to specify the desired outcome without specifying how to achieve the outcome, increases flexibility and generality. Intent-based systems also encourage innovation.

Chapter Contents

10

Orchestration: Automated Replication And Parallelism

10.1 Introduction

The previous chapter argues that the scale and complexity of a cloud data center mandates using automated systems to configure and operate equipment and software systems. The chapter also points out that many automated tools and technologies have been developed.

This chapter considers the topic of orchestration in which an automated system configures, controls, and manages all aspects of a service. The chapter examines one orchestration technology in detail. A later chapter explains the microservices programming paradigm that orchestration enables.

10.2 The Legacy Of Automating Manual Procedures

As the previous chapter points out, the easiest way to build an automation tool consists of starting with the manual processes humans use and looking for ways to have software handle some steps automatically. The cartoonist Rube Goldberg understood our tendency to automate manual processes, and made fun of the consequences in some of his cartoons. In one cartoon, for example, Goldberg shows an automatic transmission. Surprisingly, Goldberg's vehicle still has a gearshift lever. When shifting is required, a large mechanical arm extends upward from the floorboards of the vehicle, the mechanical hand grabs the gearshift lever, and the arm moves the lever exactly the way a human driver would. We laugh at such nonsensical contraptions, but there is an underlying point:

> *Automating a manual process can lead to a clumsy, inefficient solution that mimics human actions instead of taking an entirely new approach.*

10.3 Orchestration: Automation With A Larger Scope

In the case of cloud data centers, early automation systems derived from manual processes led to separation of functionality, with some tools helping deploy virtualized servers, others handling network configuration, and so on. The question arose: can designers find a way to build more comprehensive systems that cross functional boundaries and handle all aspects needed for a service? The move to containers made the question especially relevant for three reasons:

- Rapid creation
- Short lifetime
- Replication

Rapid creation. The low-overhead of containers means that it takes significantly less time to create a container than to create a VM. An automated system is needed because a human would take an intolerably long time performing the steps required to create a container.

Short lifetime. Unlike a VM that remains in place semi-permanently once created, a container is ephemeral. A container resembles an application process: a typical container is created when needed, performs one application task, and then exits.

Replication. Replication is key for containers. When demand for a particular service increases, multiple containers for the service can be created and run simultaneously, analogous to creating multiple concurrent processes to handle load. When demand for a service declines, unneeded container replicas can be terminated.

Rapid creation and termination of containers requires an automated system. In addition, the network (and possibly a storage system) must be configured quickly when a container is created or terminated. Thus, instead of merely starting container execution, an automated container management system needs a larger scope that includes communication and storage. Industry uses the term *orchestration* to refer to an automated system that coordinates the many subsystems needed to configure, deploy, operate, and monitor software systems and services.

We can summarize:

> *Unlike an automation tool that focuses on one aspect of data center operations, an orchestration system coordinates all the subsystems needed to operate a service, including deploying containers and configuring both network communication and storage.*

In addition to automated configuration and deployment, a container orchestrator usually handles three key aspects of system management:

- Dynamic scaling of services
- Coordination across multiple servers
- Resilience and automatic recovery

Dynamic scaling of services An orchestrator starts one or more copies of a container running, and then monitors demand. When demand increases, the orchestrator automatically increases the number of simultaneous copies. When demand decreases, the orchestrator reduces the number of copies, either by allowing copies to exit without replacing them or by terminating idle copies.

Coordination across multiple servers. Although multiple containers can run on a given physical server, performance suffers if too many containers execute on one server. Therefore, to manage a large-scale service, an orchestrator deploys copies of a container on multiple physical servers. The orchestrator monitors performance, and balances the load by starting new copies on lightly-loaded servers.

Resilience and automatic recovery. An orchestrator can monitor an individual container or a group of containers that provide a service. If a container fails or the containers providing a service become unreachable, the orchestrator can either restart the failed containers or switch over to a backup set, thereby guaranteeing that the service remains available at all times.

10.4 Kubernetes: An Example Container Orchestration System

Various groups have designed orchestration systems that manage the deployment and operation of services using containers. We will examine a technology that was developed at Google and later moved to open source. The technology has become popular. Known as *Kubernetes* and abbreviated *K8s*†, the technology manages many aspects of running a service‡. Figure 10.1 lists seven features of Kubernetes.

1	Service naming and discovery
2	Load balancing
3	Storage orchestration
4	Optimized container placement
5	Automated recovery
6	Management of configurations and secrets
7	Automated rollouts and rollbacks

Figure 10.1 Seven key features of Kubernetes container orchestration.

---

†Pronounced *Kates*, K8s abbreviates *Kubernetes*: the letter *K* followed by 8 letters followed by *s*

‡Proponents claim that using the term *orchestration* to describe Kubernetes is somewhat unfair because Kubernetes handles more tasks than other orchestrations technologies.

Service naming and discovery. Kubernetes allows a service to be accessed through a domain name or an IP address. Once a name or address has been assigned, applications can use the name or address to reach the container that runs the service. Typically, names and addresses are configured to be global, allowing applications running outside the data center to access the service.

Load balancing. Kubernetes does not limit a service to a single container. Instead, if traffic is high, Kubernetes can automatically create multiple copies of the container for a service, and use a *load balancer*† to divide incoming requests among the copies.

Storage orchestration. Kubernetes allows an operator to mount remote storage automatically when a container runs. The system can accommodate many types of storage, including local storage and storage from a public cloud provider.

Optimized container placement. When creating a service, an operator specifies a cluster of servers (called *nodes*) that Kubernetes can use to run containers for the service. The operator specifies the processor and memory (RAM) that each container will need. Kubernetes places containers on nodes in the cluster in a way that optimizes the use of servers.

Automated recovery. Kubernetes manages containers. After creating a container, Kubernetes does not make the container available to clients until the container is running and ready to provide service. Kubernetes automatically replaces a container that fails, and terminates a container that stops responding to a user-defined health check.

Management of configurations and secrets. Kubernetes separates management information from container images, allowing users to change the information needed for configuration and management without rebuilding container images. In addition to storing conventional configuration information, such as network and storage configurations, Kubernetes allows one to store sensitive information, such as passwords, authentication tokens, and encryption keys.

Automated rollouts and rollbacks. Kubernetes allows an operator to roll out a new version of a service at a specified rate. That is, a user can create a new version of a container image, and tell Kubernetes to start replacing running containers with the new version (i.e., repeatedly terminate an existing container and start a replacement container running the new image). More important, Kubernetes allows each new container to inherit all the resources the old container owned.

10.5 Limits On Kubernetes Scope

Although it handles many functions, Kubernetes does omit some aspects of deploying and managing containerized software. Kubernetes

- Does not focus on a specific application type
- Does not manage source code or build containers
- Does not supply event-passing middleware
- Does not handle monitoring or event logging

―――――――――――――
†See page 9 for an explanation of a load balancer.

Does not focus on a specific application type. Kubernetes is *application agnostic* in the sense that it can be used for an arbitrary type of application and an arbitrary programming paradigm. For example, it can be used to deploy stateful or stateless applications. It can be used with serverless or traditional client-server paradigms, and can support arbitrary workloads. Although it increases generality, remaining independent of the application type means that Kubernetes does not include special facilities for any specific type.

Does not manage source code or build containers. Kubernetes relies on conventional container technology to handle source code and build containers. Throughout the chapter, we will assume the use of Docker containers. Thus, when it deploys a container on a node, we will assume the node runs Linux along with the Docker software needed to run a container (i.e., dockerd).

Does not supply event-passing middleware. One programming paradigm used with containers passes asynchronous events among containers. Kubernetes does not supply the middleware needed to pass events, but instead assumes a programmer will use an extant middleware facility.

Does not handle monitoring or event logging. Kubernetes monitors and controls containers, but does not provide built-in mechanisms to collect, log, or report measurements. Some proof-of-concept systems are available for monitoring, but their use is optional.

10.6 The Kubernetes Cluster Model

Kubernetes is a sprawling technology that comprises many software components. To add further complexity, the components allow one to configure the system in a variety of ways, and various sources use inconsistent terminology when describing Kubernetes. As a result, Kubernetes can be difficult to understand, especially for beginners. One important key to understanding the software lies in starting with a conceptual model that clarifies the overall purpose and structure of the system. The model helps explain how individual software components fit into the overall picture.

Kubernetes provides *container orchestration*, which means it automates the deployment and operation of a set of one or more containers to provide a computation service. Kubernetes uses the term *cluster* to describe the set of containers plus the associated support software used to create, operate, and access the containers. The number of containers in a cluster depends on demand, and Kubernetes can increase or decrease the number as needed.

Software in a cluster can be divided into two conceptual categories: one category contains software invoked by the owner of the cluster to create and operate containers. The other category contains software invoked by users of the cluster to obtain access to a container. Figure 10.2 depicts the conceptual organization of a Kubernetes cluster, and shows the roles of an owner and users.

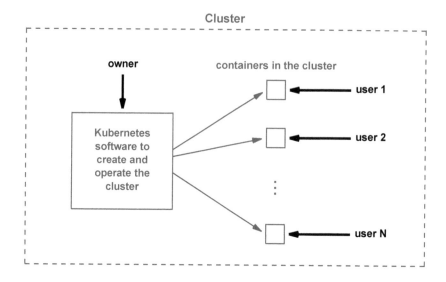

Figure 10.2 The conceptual model for Kubernetes in which an owner runs
software to create and operate a set of containers, and the containers provide a computation service that users access.

The point is:

> *When studying Kubernetes software components, it is important to
> remember that an owner invokes some software components to create
> and operate the cluster, and users invoke other components when they
> access the cluster.*

Although the terms "owner" and "user" seem to refer to humans, the roles do not
have to be filled by entering commands manually. Each of the two categories of
software provides multiple APIs. Although a human can indeed interact with the
software through a command-line interface, apps can also perform the functions (e.g.,
through a RESTful interface). In the case of a container, a user typically runs an app
that uses the traditional client-server paradigm to communicate with a container. We
will see that Kubernetes configures the network to make such access possible.

10.7 Kubernetes Pods

According to the conceptual model above, Kubernetes deploys a set of one or more
containers. In fact, Kubernetes deploys one or more running copies of a complete application program. Many applications do indeed consist of a single container. However, some programming paradigms, such as the microservices paradigm†, encourage a

---

†Chapter 12 explains the microservices paradigm.

software engineer to divide an application into small autonomous pieces that each run as a container. The pieces are tightly-coupled which means they are designed to work together, usually by communicating over the network. To run a copy of a multi-container application, all containers for the application must be started, and the network must be configured to provide communication.

Kubernetes uses the term *pod* to refer to an application. Thus, a pod can consist of a single container or multiple containers; single container pods are typical. A pod defines the smallest unit of work that Kubernetes can deploy. When it deploys an instance of a pod, Kubernetes places all containers for the pod on the same node.

In terms of networking, Kubernetes assigns an IP address to each running pod. If a pod has multiple containers, all containers in the pod share the IP address. Communication among containers in the pod occurs over the *localhost* network interface, just as if the containers in the pod were processes running on a single computer. The use of a single IP address for all containers means a programmer must be careful to avoid multiple containers attempting to use the same transport layer port. For example, if one of the containers in a pod uses port 80 for a web server, none of the other containers will be allowed to allocate port 80.

10.8 Pod Creation, Templates, And Binding Times

How does a programmer create a pod? Kubernetes uses a late binding approach in which a programmer creates a *template* for the pod (sometimes called a *pod manifest*) that specifies items to use when running the pod. A template assigns a name to the pod, specifies which container or containers to run, lists the network ports the pod will use, and specifies the version of the Kubernetes API to use with the pod. A template can use *yaml* or *json* formats; Figure 10.3 shows an example.

```
apiVersion: v1
kind: Pod
metadata:
  name: example-pod
  labels:
    purpose: web-server
spec:
  containers:
  - name: example-pod
    image: f34cd9527ae6
    ports:
    - containerPort: 8080
```

Figure 10.3 An example template for a pod that runs one container† and uses port 8080.

---

†Although the example uses a hash name for the image, *f34cd9527ae6*, it is possible to specify a human readable name.

When it deploys a pod on a node, Kubernetes stores information from the template with the running pod. Thus, changes to a template apply to any new pods that are created from the template, but have no effect on already running pods. In particular, the *metadata* specifications in the example are kept with each running pod, including *labels* that contain information useful for humans. For example, the *purpose* label has no intrinsic meaning, but a copy is stored with a pod when the pod is created from the template, meaning a running pod will have a label *purpose* with the value *web-server*. Tools allow humans to search for running pods with a label *purpose=web-server*. We will see that Kubernetes software components use template information stored with the pod to control and operate the pod.

In practice, a template will specify many more items than our example shows. A template can specify a set of environment variables and initial values to be passed to containers in the pod. A template can also specify external storage mount points for each container and an initial command to run when the container executes.

10.9 Init Containers

In addition to containers that perform the intended function of the pod, Kubernetes allows a designer to include one or more *init containers*. All init containers in a pod must complete successfully before the main containers are started. As the name implies, an init container is intended to perform initialization that might be needed before the main containers run.

One particular form of initialization is especially pertinent to containers: testing the environment to ensure that the facilities the main containers need are available. The point is to guarantee that the main containers will not encounter problems. For example, a main container may have external mount points that connect to external storage. An init container can check the external storage and either exit normally if the storage is available, or exit with an error status if the storage is unavailable. Similarly, an init container can check repositories or other computational services that containers in the pod use.

Why not incorporate checks into the main containers? There are multiple reasons, but one stands out: a given container may only require some facilities in the environment. If a container finds the facilities it needs, the container will start running. However, other containers may fail because the facilities they need are not available, leaving the pod partially running. Serializing the tests and completing them before any containers run avoids a problem in which some containers in the pod start successfully and others do not. The point is:

> *A pod designer uses a set of init containers to check the environment before the main containers of the pod execute. Doing so guarantees that either all the main containers in a pod start successfully or none of them start.*

10.10 Kubernetes Terminology: Nodes And Control Plane

Because the software is designed to run in a data center, Kubernetes uses a broad definition of a *node* to be a computational engine on which software runs. A node can be a physical server, but is likely to be a virtual machine. When an owner creates a cluster, the owner specifies a set of nodes that can be used to run the cluster, and Kubernetes spreads containers from the cluster across the nodes.

Kubernetes uses the term *Control Plane* (also known by the term *master*) to refer to the software components that an owner uses to create and operate containers. When the control plane software runs on a node, the node is known as a *master node*.

Kubernetes made an unfortunate choice of terminology by calling a node that Kubernetes uses to run containers a *Kubernetes node*. Some sources use the term *worker node*, which helps clarify the meaning. We will use the terms *master node* and *worker node* to make the purpose of each node explicit.

A typical Kubernetes cluster runs a single copy of the control plane components, and all control-plane components run on a single master node. It is possible to create a high-availability cluster by running multiple copies of the control plane software on multiple master nodes. Because such deployments represent an exception, our discussion will assume that only one copy of the control plane software runs and all the control plane software components run on a single master node.

10.11 Control Plane Software Components

The Kubernetes control plane software consists of five main software components. Figure 10.4 lists the names and purpose of each.

Component	Purpose
API server	Provides an interface to the control plane
Scheduler	Chooses a node that will run a pod
Cluster State Store	Stores state information for the cluster
Controller Manager	Handles controllers that operate the cluster
Cloud Controller Manager	Handles cloud provider interactions

Figure 10.4 The main software components that constitute the Kubernetes control plane.

API server. Sometimes labeled the *kube-api-server*, the API server provides an interface to all the control plane components. Whenever a human or a management app needs to interrogate or control a cluster, communication goes through the API server.

Scheduler. Sometimes labeled the *kube-scheduler*, the Scheduler handles the assignment of pods to nodes in the cluster. It watches for a pod that is ready to run but has not yet been assigned to a node. It then chooses a node on which to run the pod and binds the pod to the node. The Scheduler remains executing after the initial deployment of pods, which allows Kubernetes to increase the number of pods in the cluster dynamically.

Cluster State Store. The Cluster State Store holds information about the cluster, including the set of nodes available to the cluster, the pods that are running, and the nodes on which the pods are currently running. When something changes in the cluster, the Cluster Store must be updated. For example, when a pod is assigned to a node, the Scheduler records the assignment in the Cluster Store.

Kubernetes uses the *etcd* key-value technology as the underlying storage technology. Etcd is a reliable distributed system that keeps multiple copies of each key-value pair and uses a consensus algorithm to ensure that a value can be retrieved even if one copy is damaged or lost.

Controller Manager. Sometimes called the *kube-controller-manager*, the Controller Manager component consists of a daemon that remains running while the cluster exists. The daemon uses a synchronous paradigm to run a set of *control loops* that are sometimes called *controllers*†. Kubernetes includes useful controllers, such as a *Replication Controller*, *Endpoints Controller*, and *Namespace Controller*. In addition to running controllers, the Controller Manager performs housekeeping functions such as garbage collection of events, terminated pods, and unused nodes.

Cloud Controller Manager. The Cloud Controller Manager provides an interface to the underlying cloud provider system. Such an interface allows Kubernetes to request changes and to probe the underlying cloud system when errors occur. For example, Kubernetes may need to request changes to network routes or request storage system mounts. The interface also allows Kubernetes to check the status of hardware. For example, when a node stops responding, Kubernetes can check whether the cloud provider lists the node as having failed.

In addition to the five main software components listed above, Kubernetes offers a *command-line app* that allows a human to connect to the cluster and enter management commands that operate the cluster. The command-line app, which is named *kubectl*, connects to the API server.

10.12 Communication Among Control Plane Components

Recall that we assume a cluster has a single master node that runs the control plane components for the cluster. Control plane components communicate with one another, and some components communicate with outside endpoints. All internal communication goes through the API server. With the exception of the Cloud Controller Manager, which uses a provider-specific protocol to communicate with the cloud system, external communication also goes through the API server. Thus, we think of the API server as forming a central switchboard. Figure 10.5 illustrates the idea by showing communication paths among the control plane components on a master node.

---

†Chapter 13 discusses controller-based software designs and control loops.

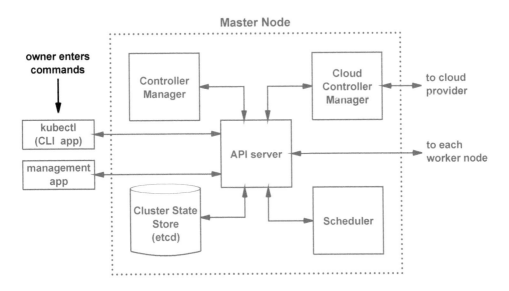

Figure 10.5 Control plane components on a master node and the communication among them.

10.13 Worker Node Software Components

Each worker node runs software components that control and manage the pods running on the node. In fact, a pod is merely an environment, so the software components manage the containers that comprise each pod. Figure 10.6 lists the main software components.

Component	Purpose
Service Proxy	Configures the network on the node (i.e., iptables)
Kubelet	Uses container software to run and monitor pods
Container Runtime	Standard container software (e.g., Docker Engine)

Figure 10.6 The main software components that run on a worker node.

Service Proxy. Sometimes called the *kube-proxy*, the Service Proxy is responsible for configuring network forwarding on the node to provide network connectivity for the pods running on the node. Specifically, the Service Proxy configures the Linux *iptables* facility.

Kubelet. The Kubelet component provides the interface between the control plane and the worker node. It contacts the API server, watches the set of pods bound to the node, and handles the details of running the pods on the node. Kubelet sets up the environment to ensure that each pod is isolated from other pods, and interfaces with the Container Runtime system to run and monitor containers. Kubelet also monitors the pods running on the node, and reports their status back to the API server.

Kubelet includes a copy of the *cAdvisor* software that collects and summarizes statistics about pods. Kubelet then exports the summary through a *Summary API*, making them available to monitoring software (e.g., *Metrics Server*).

Container Runtime. Kubernetes does not include a Container Runtime system. Instead, it uses a conventional container technology, and assumes that each node runs conventional Container Runtime software. Although Kubernetes allows other container systems to be used, most implementations use Docker Engine. When Kubernetes needs to deploy containers, Kubelet interacts with the Container Runtime system to perform the required task.

Figure 10.7 illustrates the software components that run on each worker node.

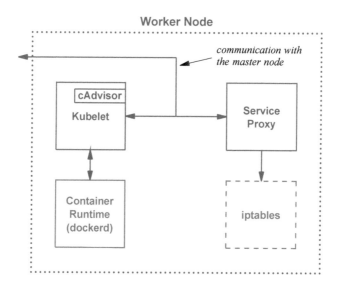

Figure 10.7 Software components on a worker node and the communication among them.

10.14 Kubernetes Features

The previous sections only cover basic concepts. Kubernetes contains many additional features and facilities, some of which are quite sophisticated and complex. The following highlights some of the most significant.

- *Replicas.* When deploying a cluster, an owner can specify and control the number of replicas for the pod. In essence, an owner can explicitly control how an application scales out to handle higher load.

- *Deployments*. Kubernetes uses the term *Deployment* to refer to a specific technology that offers an automated technology for scale out. Like other Kubernetes facilities, Deployments follow the intent-based approach. An owner specifies the desired number of replicas and the Deployment system maintains the required number of replicas automatically.

- *StatefulSets*. The StatefulSets facility allows an owner to create and manage stateful applications. A user can deploy a set of pods with guaranteed ordering. Each pod in the set is assigned a unique identity, and the system guarantees that the identity will persist.

- *DaemonSet*. The DaemonSet facility allows an owner to run a copy of a pod on all nodes in a cluster (or some of them). As the name implies, the facility is typically used to deploy a permanently-running daemon process that other containers on the node can use.

- *Garbage Collection*. Kubernetes objects have dependencies. For example, a *Replicaset* owns a set of pods. When the owner of an object terminates, the object should also be terminated (e.g., collected). The Garbage Collection facility allows one to set dependencies and specify how and when terminated objects should be collected.

- *TTL Controller for Finished Resources*. The TTL Controller allows an owner to specify a maximum time after an object has finished (either terminates normally or fails) before the object must be collected. The value is known as a *Time-To-Live* (*TTL*), which leads to the name of the facility.

- *Job*. The job facility creates a specified number of pods, and monitors their progress. When a specified number of the pods complete, the Job facility terminates the others. If all pods exit before the required number has been reached, the Job facility restarts the set. As an example, the facility can be used to guarantee that one pod runs to completion.

- *CronJob*. The CronJob facility allows an owner to schedule a job to be run periodically (e.g., every hour, every night, every weekend, or once a month). The idea and the name are derived from the Unix *cron* program, and the CronJob facility uses the same specification format as cron.

- *Services*. The Services facility allows an owner to create a set of pods and specify an access policy for the set. In essence, the Services facility hides the details of individual pods and passes each request to one of the pods automatically. Decoupling a Service from the underlying pods allows pods to exit and restart without affecting any apps that use the service. The facility works well for a *microservices* architecture, as described in Chapter 12.

For many tasks, Kubernetes offers a choice between two approaches: the ability to issue explicit, detailed commands or the use of an automated system that handles details. We can summarize:

> *Kubernetes offers many facilities, and often allows one to choose between a mechanism that offers explicit control and a mechanism that handles the task automatically.*

10.15 Summary

Instead of automating individual tasks, an orchestration system handles all the data center subsystems needed to deploy a service, including the network, storage facilities, and container execution. Orchestration typically handles dynamic scaling, coordination across multiple physical servers, and recovery.

Kubernetes, a sprawling set of software mechanisms and facilities, provides the primary example of an orchestration technology. Kubernetes handles service naming and discovery, load balancing, storage, container placement, automated container restart, management of configurations, and automated rollouts.

Kubernetes uses a cluster model in which each cluster consists of a master node that runs control plane software and worker nodes that run pods. Each pod represents an application, and may consist of one or more containers that work together on one application.

The owner of a service uses the Kubernetes control plane to deploy and control the service, The main control plane components consist of an API server, Scheduler, Cluster State Store, Controller Manager, and Cloud Controller Manager. The components communicate through the API server. External management apps also communicate with the API server; the kubectl app provides a command-line interface.

Each worker node runs a Service Proxy that configures network communication among pods on the node (i.e., configures iptables), and Kubelet that runs and monitors pods. Kubelet is not a container system, but instead uses a conventional container runtime system, such as Docker's dockerd.

Kubernetes contains many facilities that handle the tasks of deploying replicas, managing stateful applications, running a background daemon on each node, controlling the collection of terminated objects, guaranteeing a job runs to completion, running jobs periodically, and offering a service that can scale out automatically. In many cases, Kubernetes allows one to choose between a mechanism that offers explicit control and one that handles tasks automatically.

Part IV

Cloud Programming Paradigms

The Approaches Used To Create Cloud-Native Applications

Chapter Contents

11

The MapReduce Paradigm

11.1 Introduction

Previous parts of the book describe cloud infrastructure and technologies, such as VMs and containers, that can be used to create computational services. This part of the book focuses on how applications can be built that use the hardware and software facilities available in a cloud environment. It considers how cloud software can be structured as well as some of the algorithms and technologies available to software engineers.

This chapter explains the MapReduce paradigm, a widely-used approach to create applications that can scale. MapReduce is one of the earliest software designs that has been adopted for use in cloud systems. We will consider how it works and why it is especially appropriate for a cloud environment. Successive chapters in this part of the text explain other software designs used to build cloud applications, including microservices and the serverless paradigm.

11.2 Software In A Cloud Environment

The move to cloud computing has raised many questions about software. How should new software be designed for a cloud environment? Are new programming languages needed? What algorithms work best, and can better algorithms be designed for a cloud? Do the paradigms and tools used to build conventional software work for the cloud, or are new paradigms and tools needed? Will dynamically loaded libraries and related technologies work in the cloud? How should apps be structured to optimize performance in a cloud?

11.3 Cloud-Native Vs. Conventional Software

It many seem that the questions about software are moot. After all, a Virtual Machine can run a conventional operating system. If a VM in the cloud runs the same operating system as a conventional computer, most apps can be transported from the computer to the VM without being rewritten. In fact, most apps can be transferred to the VM in binary form.

> *If a VM runs the same operating system as a conventional computer, apps that run on the computer will run on the VM without being recompiled.*

Why did questions arise about the design of cloud software? The answer lies in a fundamental observation: software controls the underlying hardware, and when the underlying hardware changes, the software community has an opportunity to consider whether new software designs will make better use of the new hardware.

In the case of cloud computing, industry uses the term *cloud-native software* to refer to software that has been designed specifically for a cloud environment. The search for ways to build cloud-native software has been spurred by optimism that new software may yield significant benefits, such as:

- Improved security
- Reduced software flaws
- Enhanced performance
- Increased scalability
- Higher reliability

Improved security. Cloud infrastructure introduces entirely new security weaknesses. Perhaps new programming paradigms and tools could help software engineers avoid the weaknesses and make software more secure.

Reduced software flaws. The flaws in current software systems are well known. Perhaps new tools and paradigms could help software engineers build cloud software that has fewer flaws.

Enhanced performance. Although it is possible to port extant software to the cloud by running a VM, the software is not designed to exploit cloud facilities. Perhaps the software community can find ways to build cloud-native software that computes answers more quickly than conventional software.

Increased scalability. A cloud offers extreme parallelism. Perhaps software engineers can find ways to build systems that scale to handle large numbers of users.

Higher reliability. In addition to increasing scale, parallelism offers a chance to increase reliability. Perhaps software could arrange to perform a computation on multiple computers at the same time, using a consensus algorithm to select the final answer.

If an error occurs on one of the computers, the consensus algorithm will ignore the erroneous results, and the computation can proceed reliably.

11.4 Using Data Center Servers For Parallel Processing

One of the most significant differences between the hardware in a data center and a conventional computer lies in the amount of simultaneous processing power available. A conventional computer has a few cores, and a data center contains many physical servers that each have many cores. Thus, the hardware in a data center should be able to perform a large computation in a fraction of the time required by the hardware in a single computer. The problem must be divided into pieces, the pieces must each be assigned to a separate processor, and the results must be collected. The approach has been used quite successfully to solve large scientific problems.

Exploiting parallelism to solve a single problem requires three conceptual steps that form the basis for most parallel processing technologies.

- Partitioning
- Parallel computation
- Combination of results

Partioning. A problem must be divided into subproblems, such that each subproblem can be handled separately. For some problems, partitioning is difficult, but for others, it can be easy and fast. Many image processing problems fall into the easy category. Consider, for example, searching a high-resolution radio telescope image for a desired artifact. If N processors are available, the image can be divided into N subimages, and each can be sent to one processor. In essence, an image is a rectangular array of pixels that can be divided into several smaller rectangular arrays.

Parallel computation. In the second phase, processors work simultaneously on their part of the problem. If the subproblems can be solved independently, the processors can work simultaneously at full speed. Thus, if a single processor takes time T to *solve a problem,* N processors working simultaneously can solve the problem in time T/N. Using parallel processing in a data center seems especially exciting because a data center has thousands of processors. If a processing task takes an hour, using 60 processors could reduce the time to a minute; using 600 processors could reduce the time to less than 10 seconds.

Combination of results. In most cases, once the subproblems have been solved, the results must be combined to produce a final answer. In the case of searching a telescope image, for example, the artifact may be detected in multiple subimages. More important, the artifact may occur along a border between two subimages, causing two of the processors to report a partial discovery. Thus, combining the results requires a final processing step to check results along the borders between subimages.

Figure 11.1 illustrates the three conceptual steps used for parallel processing. The next section adds more details and explains how a software technology for parallel processing implements the steps.

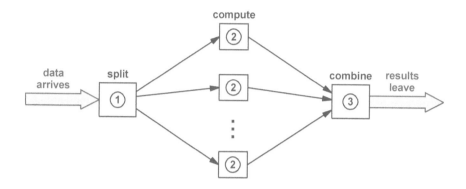

Figure 11.1 Illustration of the three conceptual steps (split, compute, combine) taken when using parallelism to solve a problem.

11.5 Tradeoffs And Limitations Of The Parallel Approach

Although it works well for some computation, the parallel approach has drawbacks that make it unsuitable for others. Five drawbacks stand out.

- Unsplittable problems
- Limited parallelism
- Resource use and cost
- Competition from databases
- Communication overhead

Unsplittable problems. The parallel approach does not work well for problems that cannot be easily partitioned. In some cases, the computation required to divide the problem into independent subproblems takes more time than parallelism saves.

Limited parallelism. Recall that when using N processors the processing time is T/N. Thus, as the number of processors increases, the time decreases. However, increasing N means dividing a problem into smaller and smaller pieces. At some point, smaller granularity becomes useless (e.g., dividing an image into individual pixels makes no sense).

Resource use and cost. Because a cloud provider charges for computational resources, using N processors may cost N times as much as one processor. Furthermore, the user's contract may place a limit on the number of simultaneous processors in use. Interestingly, in the best case, as the number of simultaneous processors increases, the time a processor is used will decrease.

Competition from databases. The question arises: is it worth building specialized software to implement the split, process, and combine steps needed for parallel processing? Researchers David DeWitt and Michael Stonebreaker have presented evidence and analysis that shows commercially-available software systems can handle many problems as well as specialized parallel systems. In particular, a modern *Database Management System (DBMS)*, which is sometimes called a *Relational Database Management System (RDBMS)*, is designed to use multiple servers to perform database operations in parallel automatically, without requiring specialized software. Moreover, many data processing problems can be handled by database operations.

Communication overhead. Sending traffic over a network may present the most significant drawback of the parallel approach. Many computations are *I/O-bound*, which means a computer spends much more time reading the input data from a file and writing output data to a file than it does performing actual computations. When splitting the problem, data must be transferred over the network to each of the N processors being used. In the worst case, the software performing the split must obtain the data from a remote storage server, and then send each piece to other storage servers. As computing proceeds, each of the processors must read its piece of the data from one of the storage servers. Consequently, the data will cross the network twice. For an I/O-bound computation, the extra I/O may mean that using a parallel approach takes longer than a non-parallel approach.

11.6 The MapReduce Programming Paradigm

Sometimes called an *algorithm*, the *MapReduce programming paradigm* extends the basic concept of parallel processing in a significant way. Instead of using a single processor to combine results, MapReduce uses parallel processing. That is, MapReduce replaces the third step in Figure 11.1 (the one labeled *combine*) with a set of processors working in parallel, with each combining *1 /K* of the results.

The data has already been split among a set of processors, so why not have each of them perform the combine step after performing the processing step? The answer is that in step two, each processor produces a set of results, but in the third step, results from all the processors must be coordinated and combined to produce the final answer. A trivial example will illustrate the combine step. After using the example to review basics, we will discuss alternative designs and tradeoffs.

Consider the problem of counting the occurrence of words in a textbook. That is, for each word that appears in the book, the output will list a count of how many times the word occurs. As one might expect, articles, prepositions, and conjunctions will occur most frequently. For example, Figure 11.2 lists the counts of the eight words that occur most frequently in one book.

Count	Word
10,500	the
9,968	to
8,553	a
5,972	at
4,851	of
4,269	from
4,055	and
3,448	with

Figure 11.2 Counts of the most-frequently occurring words in a textbook.

How can parallel processing be used to compute the frequency of words? Assume the book is divided into N files, where each file contains the text for a single chapter. One way to split the computation among multiple processors consists of assigning each chapter to a processor. The result of the computation will be N files, where each file contains a list of words in the chapter along with a count of how many times the word occurs in the chapter. For example, Figure 11.3 shows the top eight words on the lists from the first three chapters of a textbook.

Chapter 1		Chapter 2		Chapter 3	
288	the	166	the	489	the
191	a	62	to	376	a
170	and	61	a	192	to
149	of	59	of	136	of
144	to	57	and	128	and
89	that	36	that	101	is
71	is	34	in	75	that
66	in	18	is	59	an

Figure 11.3 Counts of the eight most frequently-occurring words in each of three chapters.

Unfortunately, the processors cannot compute the overall counts because each processor only computes the summary of counts from a single chapter. How can the combination step be performed in parallel? The answer lies in redistributing the chapter summaries to a set of K processors. For example, we might decide to run the combination algorithms on five parallel processors and use the first letter of each word as a way to divide words among the five. We might assign words starting with the letters A, B, C, D, and E to processor 1, words starting with F, G, H, I, and J to processor 2, and so on. Figure 11.4 lists the division.

Processor	Words Starting With
1	A through E
2	F through J
3	K through O
4	P through T
5	U through Z

Figure 11.4 One possible way to distribute counts from three chapters among five processors.

MapReduce uses the term *Map* to refer to computation performed on the first set of processors, the term *Shuffle* to refer to the redistribution of data items to a second set of processors, and the term *Reduce* to refer to the processing performed on the second set of parallel processors. Figure 11.5 illustrates the five steps involved in MapReduce processing†.

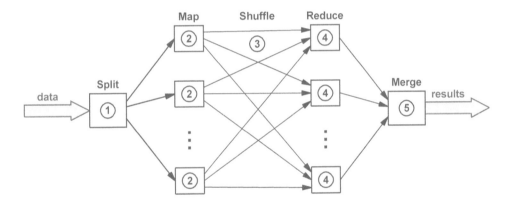

Figure 11.5 The five conceptual steps of MapReduce processing.

In the figure, data flows left-to-right through each of the labeled steps. Initially, all the data arrives at the *Split* step, which divides the data among *Map* processors. The *Shuffle* step occurs between *Map* processing and *Reduce* processing. Once the *Map* step completes on a processor, the results are divided among the processors that will perform the *Reduce* function. Thus, each *Map* processor also performs the *Shuffle* operation by choosing where to send each piece of data it produces. Similarly, once the *Reduce* step completes, the results must be merged for final output. The figure only shows a conceptual version of MapReduce; later sections discuss alternative implementations.

---

†Some sources use the term *Map* to encompass the first three steps and the term *Reduce* for the final two steps.

11.7 Mathematical Description Of MapReduce

The process is named *MapReduce* because the initial steps can described as a mathematical mapping. The Split step maps each data item, *d*, to a Map processor, m_i:

$$\text{split: d} \rightarrow \text{m}_i$$

The Map step transforms the data it receives into a list, *L*, where each item in the list consists of a pair *(K_j, V_j)*. were K_j is a *key*, and V_j is a *value*. The Shuffle step uses the keys to choose a Reduce processor, R_k for each item in the list:

$$\text{shuffle: K}_j \rightarrow \text{R}_k$$

11.8 Splitting Input

How should input be split among a set of Map processors? Several possibilities arise, and the best choice depends on the data being processed. Examples include:

- Meaningful data pieces
- Equal-sized chunks
- Hashed data values

Meaningful data pieces. Our example of a book divided into one file per chapter illustrates that data may arrive already organized into relatively large pieces that can each be sent to one Map processor. More important, for some data, it may be necessary to send an entire logical piece of data to a given Map processor because the Map computation needs to examine the entire piece.

Equal-sized chunks. Using meaningful pieces of data may not produce optimal results. For example, dividing a book by chapters means that parallelism is limited to the number of chapters. Furthermore, the meaningful pieces of data may not be the same size, so the Map computation will not be divided equally among the processors. If some Map computations take much longer than others, parallelism does not pay off. Thus, an alternative divides the input data into N equal-size chunks, and sends one chunk to each Map processor. For example, to equalize processing for our book example, we could concatenate all chapters of the book, treat the result as lines of text, and send *1/N* lines to each Map processor.

Hashed data values. An especially useful technique uses a hash function to map data items into integers *1* through *N* that each specify a Map processor. If the data arrives with an identifying key on each record, it may be possible to compute the hash from each key. Even if the data is unstructured, a hash can be computed on chunks of data.

11.9 Parallelism And Data Size

For our trivial example of counting words in a book, the use of parallel processing may seem irrelevant. Indeed, the computation can easily be handled by a single computer. The use of parallelism and a MapReduce approach merely represents unwarranted overhead. In many large problems, however, using parallelism can be justified because the problem involves a huge amount of data and both the Map and Reduce steps involve complex computation.

How large must a problem be to justify MapReduce? There is no easy answer. Some sites recommend splitting the input into blocks of at least 64 Megabytes; the input should consist of many such blocks. Others recommend that one Terabyte is the minimum size. Hyperscaler providers have used MapReduce with Petabytes of data. Of course, one cannot use the size of data alone to determine whether to use MapReduce. The amount of computation required for the Map and Reduce steps must be considered as well as the importance of obtaining results quickly and the cost. The point is:

> Programmers should remain aware that MapReduce is only intended for problems that involve complex processing on extremely large volumes of data. When applied to smaller problems, MapReduce merely introduces unnecessary overhead.

11.10 Data Access And Data Transmission

The diagram of MapReduce in Figure 11.5† implies that data moves processor to processor. In practice, transferring data directly may not be efficient or necessary. To see why, consider splitting the input data into N equal-size blocks. From the figure, it may seem that the Split computation accepts all the input data, places it in local storage, and once all data has arrived, calculates the size of blocks, and sends each block to a Map processor. A more efficient approach obtains the size of the data from the underlying storage system without actually reading the data. Furthermore, if the Map processors can each access the storage system directly, the only data that needs to pass from the Split computation to a Map processor consists of an object name and a range of bytes in that object. Of course, placing processing near the data can also reduce the load on the network; we will see that tools are available to help optimize data transfer.

Only the output from a Map computation is passed on through the Shuffle to the Reduce processors. Thus, once a Map computation finishes processing, the data can be discarded. Typically, a Map computation places intermediate results in a local file. The contents can then be sent through the Shuffle. As an alternative, it is possible to arrange for the Map computation to write data to a global store, and then arrange for the Reduce processors to access the store. However, doing so may require two transfers across the data network instead of one.

---

†Figure 11.5 can be found on page 151.

Our textbook example illustrates an interesting point about data sizes: the output from the Map computation may be much smaller than the original data. The text for the chapters (with typesetting commands removed) comprises 1,006,956 bytes. However, the list of words in the text along with their counts takes only 104,975 bytes, about 10% of the original data size. Thus, when arranging a data transmission scheme, a programmer must consider the size of the mapped data as well the size of the original data.

> *When using MapReduce, a programmer must consider data transmission. Mapped data can be much smaller than the input data. If the input data resides on a global storage system, data copying can be reduced by arranging for Map processors to access the data directly rather than arranging for the Split computation to access the data and send pieces to each Map processor.*

11.11 Apache Hadoop

The idea of MapReduce was pioneered at Google, where it is used to process Web data and generate results for search queries. It was popularized by *Apache Hadoop†*, an open source project that has produced a set of tools that can be used to implement MapReduce. Social media and search companies, including Google, Facebook, Yahoo!, and Amazon, use Hadoop, as do scientists and academics.

Like many open source projects, Hadoop has received contributions from many groups and individuals. The software has evolved through three major versions, and has grown to gargantuan size. In addition to the basic system, auxiliary tools and enhancements are available that increase performance and reliability. Third-party software is also available that extends various aspects of Hadoop. Consequently, our brief description of Hadoop only covers the basics.

11.12 The Two Major Parts Of Hadoop

Hadoop software can be divided into two major parts that work together:

- Processing component
- Distributed file system

Processing component. Hadoop's processing component provides software that automates MapReduce processing. To use Hadoop, a user must build a piece of software to perform the Map function and a piece of software to perform the Reduce function. The user must also configure Hadoop. The processing component contains software that uses the configuration information to deploy copies of the Map code on multiple processors, split the data among them, configure network addresses for the shuffle step, and deploy copies of the Reduce code. In addition, Hadoop can handle reliability by

---

†One of the original Hadoop designers took the name from their child's toy elephant.

monitoring each of the elements and restarting items that fail. Hadoop can also monitor processing and restart computation after a failure. In other words, once a user writes the two basic pieces of software, Hadoop automates most of the configuration and operation.

Distributed file system. Recall that moving data from storage to Map processors and moving mapped data to Reduce processors can form significant sources of overhead when using the MapReduce paradigm. Hadoop addresses the problem directly by defining its own distributed file system that holds the data being processed. Known as the *Hadoop Distributed File System* (*HDFS*), the system provides high-speed access tailored to the specific requirements of MapReduce. To understand how Hadoop optimizes processing, one needs to understand the Hadoop infrastructure. The next section explains the model.

11.13 Hadoop Hardware Cluster Model

To understand the motivation for Hadoop's distributed file system one must understand the hardware system for which Hadoop was designed. Hadoop uses the term *cluster* to refer to a set of interconnected computers that run Hadoop. In the original configuration, each computer in the cluster had a local disk. Figure 11.6 illustrates cluster hardware.

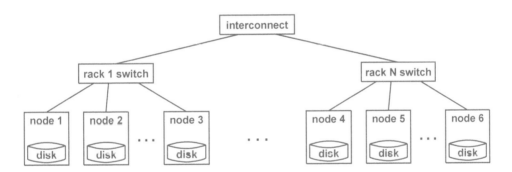

Figure 11.6 Illustration of a dedicated Hadoop cluster consisting of racks of nodes, where each node contains a local disk.

To achieve high degrees of parallelism, a Hadoop cluster needs many nodes. A typical cluster contains between 1,000 and 4,000 nodes with approximately 40 nodes per rack. In terms of node capabilities, because MapReduce processing focuses on data, disk space is often more important than processing power. For example, Facebook reported a cluster in which each node had eight 1.5 Terabyte disks (individual disks rather than a RAID array). However, a node in the Facebook cluster only had 16 cores and 32 Gigabytes of RAM.

11.14 HDFS Components: DataNodes And A NameNode

HDFS divides each file into fixed-size blocks of 128 Megabytes per block, and defines two types of nodes used to store files. A set of *DataNodes* each holds blocks from the files being stored. For example, a DataNode with 12 Terabytes of disk space can store up to 98,304 blocks. A single *NameNode* provides the binding between a file name and the location of blocks in the file (i.e., DataNodes on which blocks of the file have been stored). Conceptually, the NameNode contains a table for each file. The table has an entry for each block of the file, and the entry specifies one or more DataNodes where copies of the block have been stored.

11.15 Block Replication And Fault Tolerance

A given HDFS has exactly one NameNode, which means the overall file system has a single point of failure. If the NameNode fails or becomes unreachable, the entire file system becomes unusable. Various attempts have been made to increase the system's fault tolerance, including saving snapshots of the NameNode for use in recovery after failure and a *High Availability* version of Hadoop that has a redundant NameNode.

Despite having only one NameNode, HDFS provides redundancy for the data stored in files. When storing a data block, HDFS creates redundant copies and distributes the copies across multiple nodes. Typically, HDFS stores three copies of each data block. Thus, each entry in NameNode contains a list of the three DataNodes on which copies of the specified block have been stored. Figure 11.7 shows a table for file *X* in a NameNode and four DataNodes holding the data blocks of the file.

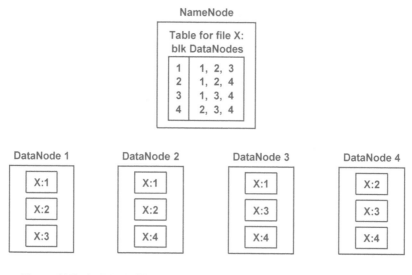

Figure 11.7 An HDFS file named *X* with three replicas of each block stored in DataNodes and a table for the file stored in the NameNode.

In the figure, the first entry in the table indicates that copies of block 1 from the file have been stored in DataNodes 1, 2, and 3. Similarly, copies of block 3 can be found on DataNodes 1, 3, and 4. Because each DataNode has blocks from multiple files, the figure labels each block with the file name X and a block number.

Hadoop is *topology aware*, which means that the software understands the interconnections among hardware units in the cluster. In particular, HDFS understands that nodes in a given rack receive power from the same source and connect to the same Top-of-Rack switch. Thus, when it distributes redundant copies of a block, HDFS uses the topology to choose locations that maximize fault tolerance.

11.16 HDFS And MapReduce

The Hadoop design links the MapReduce computation and HDFS. Three aspects stand out:

- Optimal split size
- File semantics optimized for MapReduce
- Colocation of computation

Optimal split size. Knowing that the underlying file system stores large, fixed-size blocks allows a Hadoop user to optimize performance. In particular, because data will be stored in HDFS blocks, a programmer can optimize performance by splitting the input data into chunks that each fit into a single block. Of course, it may not be possible to divide input data on arbitrary byte boundaries. For example, the input may consist of records, Map processing may need to operate on an entire record at a time, and the size of an HDFS block may not be an exact multiple of the record size. Fortunately, for most applications, the size of a record is much smaller than the size of an HDFS block, meaning that a block can hold many input records. As a result, very little storage space is wasted by placing full records in each block and leaving any remainder of the block empty. The point is:

> *When using Hadoop with an HDFS file system, a programmer can optimize performance by choosing to split data into chunks that each fit into a single HDFS block.*

File semantics optimized for MapReduce. Interestingly, HDFS is not a general-purpose file system, and it does not provide the usual set of file operations. Instead, HDFS only offers two basic data operations: sequential *read* and *append*. That is, when an application *reads* from an HDFS file, the software fetches successive blocks of the file as needed. Similarly, when an application adds data to the end of a file, HDFS creates new blocks as data is generated, places copies of each block on multiple DataNodes, and updates the table in the NameNode.

The limited functionality suffices because the MapReduce paradigm only needs the two operations. When it splits input data, MapReduce generates each split sequentially. When it processes a given subset of the data, a Map step reads the data sequentially, and so on. None of the steps needs random file access, and none of the steps alters existing files. The limited functionality allows HDFS to be optimized. We can summarize:

> *HDFS is optimized for the file operations that MapReduce needs: sequential* read *and the ability to* append data onto a file.

Colocation of computation. Perhaps the most significant link between HDFS and MapReduce processing occurs because Hadoop locates MapReduce processing on the same nodes that store the data. In particular, Hadoop attempts to perform Map processing on the DataNodes that hold the data to be processed. The point is:

> *Hadoop achieves high performance by scheduling each computation on the same nodes that hold the data for the computation.*

11.17 Using Hadoop With Other File Systems

Recall that Hadoop was designed to work with HDFS and run on a hardware cluster in which each node has a local disk. Can Hadoop be run in a cloud data center that does not have local storage on each node? The short answer is yes.

Hadoop has been used with many file systems, including NAS and object stores. The disadvantage of using remote storage lies in transmission overhead, as described previously. The Split step must move data from remote storage to the processor performing the split, and the pieces must be written back to remote storage. Then, the Map software must read pieces and process them. Fortunately, modern data center networks have sufficient capacity to handle large volumes of data. Furthermore, distributed object storage technology can access data blocks faster than a traditional file system. Thus, Hadoop can still exhibit reasonable performance.

11.18 Using Hadoop For MapReduce Computations

How can one use Hadoop? Two basic approaches are possible:

- Manual setup and configuration
- Cloud provider facilities

Manual setup and configuration. Hadoop software is freely available. Downloads of experimental versions as well as stable production versions can be obtained from:

hadoop.apache.org

Hadoop offers a programmer a way to experiment with Hadoop software on a single node cluster (i.e., on a single computer). The idea is to build and try Map and Reduce software and learn how to manage Hadoop before attempting to create and operate a large, production deployment. The following web page contains a link that explains how to experiment with a single-node cluster:

https://hadoop.apache.org/docs/stable/

Cloud provider facilities. When running Hadoop in a public cloud, a programmer can follow one of two approaches. In one approach, a programmer leases a set of VMs, downloads the Hadoop software, and deploys a cluster manually. In the other approach, a programmer uses a special interface that the provider offers. The interface allows a programmer to specify parameters for a Hadoop cluster, and invokes an automated system that configures the cluster. Thus, a programmer can use Hadoop without setting up and managing the software manually.

Amazon's *Elastic Reduce* provides an example of MapReduce software offered by a cloud provider. The system offers both web and command-line interfaces, runs MapReduce jobs on Amazon's EC2, stores data in Amazon's S3, and includes monitoring that can shut down machines when processing completes.

11.19 Hadoop's Support For Programming Languages

Hadoop software includes interfaces that support multiple programming languages. That is, a programmer who decides to write Map and Reduce functions can choose a programming language, which can differ from the language in which Hadoop is implemented. Popular language interfaces include:

- Hadoop's Java API for Java programs
- Hadoop's Streaming mechanism for scripting languages (e.g., Python)
- Hadoop's Pipe API for use with C and C++

The key idea is that Hadoop gives a programmer the freedom to choose a language that is appropriate to the problem being solved:

> *Unlike software systems that restrict programmers to one programming language, Hadoop offers mechanisms that give a programmer a choice among popular languages.*

11.20 Summary

Although VMs allow arbitrary applications to be transported to a cloud data center, building cloud-native software offers a chance for improved security, reduced flaws, enhanced performance, increased scalability, and higher reliability.

One particular style of computation is especially appropriate for a cloud data center: parallel processing. The basis for using parallel computers to increase performance lies in three steps: partioning a problem into pieces, using multiple computers to process the pieces at the same time, and combining the results. Parallel processing does have potential limitations, including unsplittable problems, limited parallelism, limited resources and cost, competition from databases, and communication overhead (especially for I/O-bound computations).

The MapReduce programming paradigm (sometimes called the MapReduce algorithm) offers a formalized framework for parallel processing. MapReduce divides processing into five steps: Split, Map, Shuffle, Reduce, and Merge. Depending on the problem being solved, data can be split into meaningful pieces, fixed-size chunks, or by using a hash of the data items.

MapReduce works best for large, complex problems. For problems that process a small amount of data or for which the computation is trivial, MapReduce introduces unnecessary overhead. A programmer must also consider data transmission because sending data takes time. Data copying can be reduced by arranging for the processors that perform Map operations to access data directly.

The Apache Hadoop project has produced a set of software tools that automates many of the mundane steps needed to implement MapReduce. Hadoop software can be divided into two major parts: a processing component and the Hadoop Distributed File System, HDFS. The file system is designed to work on a hardware cluster in which each node contains a local disk. HDFS uses a set of DataNodes to hold blocks of files and a NameNode to hold metadata that specifies where the blocks for a given file have been stored.

When using HDFS, a programmer can optimize performance by splitting data into pieces that each fit into a single HDFS block. HDFS operations are optimized for MapReduce processing; only *read* and *append* operations are supported. Hadoop achieves a higher performance by scheduling computation on the nodes where data resides. Although Hadoop can be used with other systems other than HDFS, performance may suffer.

A programmer can download Hadoop software, install the software, and manage it manually; starting with a single node cluster allows a programmer to become familiar with Hadoop before deploying a large cluster. As an alternative, a programmer can use a Hadoop facility offered by a cloud provider. Such facilities have web and command-line interfaces that relieve a programmer from having to install and manage Hadoop software.

Unlike some tools, Hadoop gives a programmer freedom to choose a programming language in which to write code. In particular, a programmer is not restricted to use the same language that is used to implement Hadoop.

Chapter Contents

12   Microservices

12

Microservices

12.1 Introduction

The previous chapter describes the MapReduce paradigm that employs parallel computation to solve a single problem. The chapter explains how splitting input data into pieces allows computation to proceed on the pieces simultaneously.

This chapter introduces an alternative way to structure software that takes advantage of multiple computers in a data center to scale computation. Instead of focusing on one problem, however, the approach divides each application into pieces, allowing the pieces to scale independently, as needed.

12.2 Traditional Monolithic Applications

How should an application program be structured? Software engineers have grappled with the question for decades. One of the key aspects arises from the question of whether to build each application as a single, self-contained program or to divide applications into multiple pieces that work together. Each approach has advantages and disadvantages.

Traditionally, software vendors favor the self-contained approach. Bundling everything into a single executable means a user only has one piece of software to install, configure, and use. Bundling eliminates failures that arise if a user fails to install all the needed pieces as well as incompatibility problems that can arise if a user upgrades or reconfigures one piece without upgrading or reconfiguring the others. Basically, the *settings* or *preferences* menu shows the user all possible choices and ensures that the entire program follows the user's selection.

Industry uses the term *monolithic* to characterize an application constructed as a single, self-contained piece of software. Monolithic applications tend to be both large and complex. Internally, they contain code to perform many functions that can span a wide range.

As an example of a monolithic application, consider an application program that provides an online shopping service. To access the service, a user launches a browser and enters the URL of the service, which connects to the application in question. The application allows the user to select items, specify shipping, enter payment information, and rate the experience. Internally, a monolithic application must contain the code needed to handle each of the steps, including code to search the catalog of products, code to access the customer database, the payment processing system, and so on. The code is organized into a main program plus a set of *functions* (which are known as *methods*). Figure 12.1 illustrates the idea.

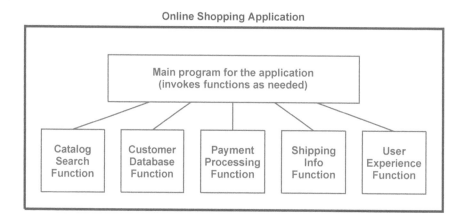

Figure 12.1 Illustration of a monolithic application with all the functions needed to support online shopping built into a single program.

12.3 Monolithic Applications In A Data Center

Can a traditional monolithic application run in a data center? Yes. A tenant can launch and use a VM to run monolithic applications, exactly the way they run on a traditional server. However, software engineers have pointed out that the monolithic approach has some disadvantages and an alternate approach may be better for a cloud environment. In particular, monolithic applications cannot be replicated as quickly as cloud-native applications. First, starting a VM has higher overhead than starting a container. Second, a monolithic design means all code must be downloaded when the application starts, even if pieces are not used.

To understand why downloading code causes unnecessary overhead, consider the shopping example and remember that many cloud systems handle scale by creating copies of an application as needed and allowing the copies to expire when they complete. In the case of online shopping, many users peruse a catalog without logging in, without making a purchase or arranging shipping, and without filling out the experience questionnaire. For such customers, the application only invokes the catalog search function; code for other functions must be downloaded, but is not used.

12.4 The Microservices Approach

The *microservices* approach to software, sometimes called a *microservices architecture*, divides functionality into multiple, independent applications. Each of the independent applications is much smaller than a monolithic program, and only handles one function. To perform a task, the independent applications communicate over a network.

The microservices approach can be used in two ways: to implement a new application or to divide an existing monolithic application. We use the term *disaggregation* to refer to the division of a monolithic application into microservices. Figure 12.2 illustrates one possible way to disaggregate the monolithic online shopping application from Figure 12.1.

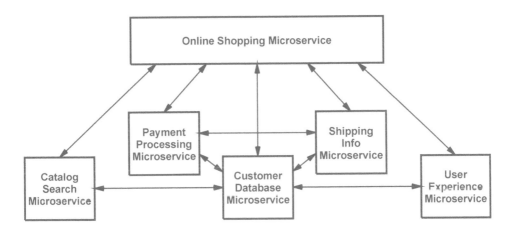

Figure 12.2 Illustration of the shopping application disaggregated into a set of microservices that communicate with one another.

12.5 The Advantages Of Microservices

It may seem that the microservices approach merely introduces the extra overhead of running multiple, small applications and using network communication among the pieces instead of internal function invocation. In a cloud environment, however, the microservices approach has advantages that can outweigh the overhead. The advantages can be divided into two broad categories: advantages for software development and advantages for operations and maintenance.

12.5.1 Advantages For Software Development

The microservices approach offers several advantages for software development:

- Smaller scope and better modularity
- Smaller teams
- Less complexity
- Choice of programming language
- More extensive testing

Smaller scope and better modularity. The microservices approach means software engineers focus on one small piece of the problem at a time and define clean interfaces. The limited scope encourages better decomposition and allows engineers to understand each piece completely. As a result, engineers are less likely to make mistakes or over-look corner cases.

Smaller teams. As the old saying goes, "too many cooks spoil the broth."   When a large number of software engineers attempt to build a large piece of software, they must all agree on many details. Because a microservice can be designed and implemented in-dependent of other microservices, each microservice only requires a small development team, meaning that the resulting code will be more uniform and less prone to errors.

Less complexity. Complexity leads to errors, and the monolithic approach creates complexity. For example, a monolithic design allows global variables to be shared among all pieces of code. By contrast, the microservices approach eliminates global variables, and requires designers to document the exact interfaces among pieces.

Choice of programming language. When using the monolithic approach, all code must be written in a single programming language. With the microservices approach, software engineers can choose the best language for each service.

More extensive testing. Testing and debugging form a key part of the software development process. Unfortunately, testing a monolithic program poses a challenge because the pieces can interact. Thus, many combinations of inputs must be used to guarantee that interactions among the pieces do not result in errors. For the largest ap-plications, the number of possible combinations is so large that engineers can only use randomized testing. With the microservices approach, each service can be tested in-dependently, allowing more extensive and thorough assessment.

12.5.2 Advantages For Operations And Maintenance

The microservices approach also offers operational advantages:

- Rapid deployment
- Improved fault isolation
- Better control of scaling
- Compatibility with containers and orchestration systems
- Independent upgrade of each service

Rapid deployment. Because each service is small, the microservices approach means a given microservice can be created, tested, and deployed rapidly. Thus, the implementation of a microservice can be changed easily and quickly.

Improved fault isolation. Dividing an application into multiple microservices makes fault isolation easier. When a problem occurs, a manager can identify and test the misbehaving microservice while allowing applications and other microservices to continue normal operations.

Better control of scaling. Each microservice can be scaled independently. In our trivial example, the microservices are only used by one application. In practice, if multiple applications use a given service, the microservice can be scaled to handle the load without scaling other services that are not heavily used.

Compatibility with containers and orchestration systems. A key distinction between the monolithic approach and the microservices approach arises from the underlying platforms needed. Because it is small and only performs one task, a microservice fits best into the container paradigm. Furthermore, using containers means that microservices can be monitored, scaled, and load balanced by a conventional container orchestration system, such as Kubernetes.

Independent upgrade of each service. Once an improved version of a microservice has been created, the new version can be introduced without stopping existing applications and without disturbing other microservices. Moreover, a manager can change a microservice without needing to recompile and replace entire monolithic applications that use the service.

12.6 The Potential Disadvantages Of Microservices

Although it offers many advantages, the microservices approach also has potential disadvantages, including:

- Cascading errors
- Duplication of functionality and overlap
- Management complexity
- Replication of data and transmission overhead
- Increased security attack surface
- Workforce training

Cascading errors. One of the advantages of a monolithic application lies in the property of being self-contained. If a function contains an error, the application may fail, but other applications will continue. In the microservices approach, one microservice can invoke another, which can invoke another, and so on. If one of the microservices fails, the failure may affect many others as well as the applications that use them.

Duplication of functionality and overlap. Interestingly, the ease with which microservices can be created invites software engineers to deploy many. For example, some large enterprises report having hundreds of microservices deployed. When functionality is needed that differs slightly from an existing microservice, it is often easier to create a completely new one than to modify the existing microservice.

Management complexity. Each microservice must be monitored and managed. Microservices make management more complex. Although orchestration and automation tools help, with hundreds of microservices running simultaneously, it can be difficult for a manager to understand their behaviors, interdependencies, and the interactions among them.

Replication of data and transmission overhead. Recall that the monolithic approach allows functions to share global data, Although it can lead to problems, sharing has the advantage of efficiency, especially in cases where functions access large sets of data. The microservices approach requires each microservice to obtain a copy of the needed data, either from a storage server or by being passed a copy when the microservice is invoked. More important, an application that invokes many microservices may experience excessive overhead if each of the microservices needs to access the data.

Increase security attack surface. We use the term *security attack surface* to refer to the points in a system that can be used to launch an attack. A monolithic application represents a single attack point. By disaggregating an application into many independent pieces, the microservices approach creates multiple points that an attacker can try to exploit.

Workforce training. When following the monolithic approach, a software engineer divides the code into modules that make the application easy to create and maintain. In contrast, the microservices approach requires software engineers to consider the cost of running each microservice as well as the communication costs incurred in accessing data or passing it from one microservice to another. A software engineer must also consider the question of granularity, as the next section explains. Thus, software engineers need new skills to create software for microservices.

Clearly, someone who creates a microservice must weigh the disadvantages as well as the advantages. A designer must make careful choices and consider the potential consequences of each. The next sections consider some of the choices and tradeoffs.

12.7 Microservices Granularity

How much functionality should be packed into each microservice? The question of microservice size forms one of the key decisions a software engineer faces when following the microservices approach. As an example of granularity choices, consider the example structure for the online shopping service in Figure 12.2†. Although the figure shows a single microservice for payment processing, other designs are possible. Payment processing may involve:

- Applying customer discounts or promotional codes
- Splitting charges (e.g., among multiple credit cards)
- Authorizing the customer's credit card(s)
- Authorizing the customer's PayPal account

---

†Figure 12.2 can be found on page 165.

Should a separate microservice be used for each payment processing function or should they all be collected into a single microservice? If separate microservices are used, should the design have a payment processing microservice that calls the others, or should the online shopping microservice call them? Figure 12.3 illustrates two alternative designs.

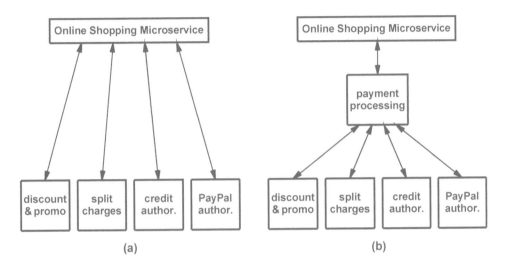

(a)   (b)

Figure 12.3 Alternatives for the payment microservice from Figure 12.2. (a) divides the functionality into four separate microservices, and (b) places the four under an intermediate microservice.

The figure shows two ways to structure the payment microservices and does not show other microservices needed for the application. In particular, additional communication will be required as shown in Figure 12.2 (e.g., between each of the microservices and the customer database microservice as well as communication with the shipping information microservice).

We know that separate microservices have advantages, such as the ability to scale independently. However, adding microservices introduces more management complexity and increases both communication and the security attack surface. What is the best way to structure the payment processing example?

There is no easy answer. However, three heuristics can help developers choose a granularity:

- Business process modeling
- Identification of common functionality
- Adaptive resizing and restructuring

Business process modeling. One rule-of-thumb suggests that each microservice should be based on a business process. Instead of merely disaggregating existing applications, development teams are encouraged to ask how the applications are being used. What are the steps along the work flow? Once they have been identified, the steps can be transformed into microservices.

Identification of common functionality. A second rule-of-thumb follows good software engineering practice: when designing a piece of software, consider the pieces of software that can use it, and ensure it satisfies all of them. The idea applies directly to microservice design: instead of building a microservice for exactly one application, consider how related applications might use the service and plan accordingly. For example, a payment processing microservice might be used by a microservice that bills wholesale customers as well as a microservice for online retail shopping.

Adaptive resizing and restructuring. The small size of a microservice means it can be redesigned quickly. The third rule-of-thumb takes advantage of the size: iterate to find the best size and structure for a microservice. In most cases, software engineers will make reasonable choices initially. However, requirements can change. Suppose, for example, an organization decides to deploy a new application that has minor overlap with existing applications. Creating an entirely new set of microservices for the new application may lead to unnecessary duplication, and thereby, increase management complexity unnecessarily. It may be better to restructure existing microservices to make them useful for both existing and new applications. One restructuring approach separates the underlying functionality from the interface, allowing the interface and the underlying functionality to change independently.

As an example of adaptive restructuring, consider a new application that uses a Blockchain payment technology. Figure 12.4 illustrates how the design in Figure 12.3(b) can be restructured to accommodate the new application.

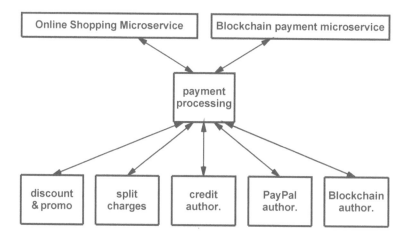

Figure 12.4 A restructured version of Figure 12.3(b) that accommodates a new application and a new payment type.

As the figure shows, a Blockchain authorization microservice has been added. In addition, the interface offered by the intermediate payment processing microservice must be extended to accommodate requests for Blockchain payments. The payment processing microservice must also be modified to communicate with the Blockchain microservice. However, the interface remains backward-compatible, which allows requests from the online shopping microservice to work exactly as before.

We can summarize:

> *Three heuristics help designers choose a granularity for microservices: model a business process, identify functionality common to multiple applications, and adaptively resize and restructure as needed. The small size of microservices makes restructuring easy and accommodates new applications and new underlying functionality.*

12.8 Communication Protocols Used For Microservices

Two questions arise about the communication among microservices, and we will see that the two are related:

- What communication protocols do microservices use?
- What communication paradigms do microservices follow?

This section examines communication protocols; the next section discusses common paradigms that microservices use.

To have meaningful and unambiguous communication, both sides must agree on the format and meaning of messages that are exchanged; communication protocols specify the message details. Like other applications in a data center, microservices communicate using Internet protocols. Doing so means a microservice can be reached from inside or outside the data center, subject to security restrictions. For the *transport layer protocol*, most microservices use the *Transmission Control Protocol* (*TCP*), with TCP being sent in *Internet Protocol* (*IP*) packets.

TCP merely delivers streams of bytes between a pair of communicating entities. When communicating over TCP, the pair must also employ a *transfer protocol* that defines how bytes are organized into messages. In essence, the set of transfer protocol messages defines the service being offered.

Although a software engineer can design a special-purpose transfer protocol for each new microservice, most engineers choose from among existing transfer protocols. Using an existing protocol makes it easier to write code. Several transfer protocols exist and satisfy most needs. Thus, a software engineer merely needs to choose one when designing a microservice. We will consider two examples:

- HTTP – The *HyperText Transfer Protocol* used in the Web
- gRPC – An open source† high-performance, universal RPC framework

---

†gRPC was originally created at Google and then moved to open source.

HTTP. When an entity uses HTTP to communicate with a microservice, data can flow in either direction. That is, the entity can send data to the microservice or request that the microservice send data. To communicate, the entity sends one or more request messages to which the microservice responds. In addition to specifying an operation to be performed, each request message specifies a data item by giving the item's name in the form of a *Uniform Resource Identifier* (*URI*). For some operations, the sender must also supply data to be used for the request. Figure 12.5 lists six basic operations that HTTP supports.

Operation	Meaning
GET	Retrieve a copy of the data item specified in the request
HEAD	Retrieve metadata for the data item specified in the request (e.g., the time the item was last modified)
PUT	Replace the specified data item with the data sent with the request
POST	Append the data sent with the request onto the specified data item
PATCH	Use data sent with the request to modify part of the data item specified in the request
DELETE	Remove the data item specified in the request

Figure 12.5 Six basic operations that can be used in HTTP messages.

gRPC. Unlike most transfer protocols, gRPC does not define a specific set of operations that can be performed. Instead, it provides a general framework for communication and allows a specific set of operations to be defined for each instance (i.e., for each microservice).

To understand gRPC, one must know that the *Remote Procedure Call* (*RPC*) approach has been used to build distributed systems for decades. The general idea is straightforward: create a program that runs on multiple computers by placing one or more of the procedures† from the program on remote computers. Instead of using normal, internal procedure calls to invoke the remote procedures, the program must be built with code that sends a message to the correct remote procedure. The message contains arguments for the procedure being called, and a reply message contains the value returned by the procedure.

To make RPC easy to use, technologies exist that generate message passing code automatically. A software engineer merely needs to specify the types of arguments and the return value for a remote procedure. The RPC system generates all the code needed

---

†The term *procedure* is an older term; synonyms include *function*, *method*, or *subroutine*.

to gather arguments, form a message, send the message to the remote procedure, receive a reply, and send the results back to the caller. RPC technologies operate transparently, meaning that they can be used to transform a conventional program merely by inserting new code —existing procedure calls in the code remain unchanged and existing procedures remain unchanged. The point is:

> *RPC technologies allow one to create and test a conventional program and then generate code that allows some of the procedures to be placed on a remote computer.*

To achieve transparency, RPC technologies generate code known as *stubs*. In the program, a stub replaces each procedure that has been moved to a remote computer; on the remote computer, a stub calls the procedure with a local procedure call, exactly like a program does. Figure 12.6 illustrates the idea. Although the example in the figure shows both procedures running on the same remote computer, RPC technology allows each remote procedure to run on a separate computer.

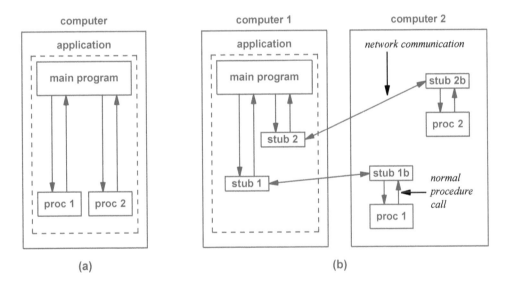

Figure 12.6 (a) An application program that calls two procedures, and (b) the same application using RPC technology that allows the procedures to run remotely.

In part (a) of the figure, the vertical arrows leading from the main program to each of the procedures represent a procedure call and return. In part (b), the main program still makes normal procedure calls. However, stub code has been substituted that communicates over a network to a remote computer. On the remote computer, a remote stub receives communication and places a normal call to the procedure. When the procedure call returns, the remote stub sends the results back to the corresponding stub on computer 1 which returns the results to the main program. The key idea is that each stub on computer 1 mimics an actual procedure and each stub on computer 2 mimics a main program. Thus, neither the main program nor the procedures need to be changed.

gRPC follows the traditional RPC approach of using stubs, but extends the technology in two ways. First, unlike most RPC systems, gRPC allows a choice among programming languages, including C, C++, Java, Go, Node.js, Python, and Ruby. Second, unlike most RPC systems, gRPC does not specify a data format to be used over the network between stubs (e.g., JSON). Instead, gRPC includes a technology known as *protocol buffers* that allows a software engineer to build software that serializes data for transmission. Unlike HTTP, which requires a separate request for each data item, protocol buffers provide a way to stream multiple data items across the network connection. The point is:

> *gRPC extends traditional RPC technology by supporting many programming languages, allowing the user to define data serialization, and supporting streaming of multiple data items.*

12.9 Communication Among Microservices

A variety of interactions have been used with microservices. The interactions can be divided into two broad types:

- Request-response (REST interface)
- Data streaming (continuous interface)

Request-response. Software engineers use the term REST or RESTful† to describe the *request-response* style of interaction used on the Web: a web browser sends a request to which a web server responds. A microservice that uses request-response interaction follows the same pattern of accepting a request and returning a response. Any further interaction between the microservice and the entity using the microservice requires another request. Saying that a microservice uses a REST API usually implies that the microservice uses HTTP as its transfer protocol.

Data streaming. A major disadvantage of request-response interaction arises in cases when a microservice needs to return many items in response to a request. To adhere to the request-response approach, the entity using the microservice must make repeated requests. Of course, a microservice can combine a small number of items into a single response (e.g., by sending the equivalent of a zip file). However, all items

---

†Although REST expands to *Representational state transfer*, most software engineers think of the expansion as a cute term contrived to give a new name to the interaction used on the Web since its inception.

must be available, and the processing required to create the combined response must be reasonable. Data streaming interaction avoids repeated requests by allowing the microservice to provide a stream of data items as a response. When using a streaming interface, an entity establishes a network connection with the microservice and sends a single request. The microservice sends a sequence of one or more data items in response to the request (i.e., the microservice streams a sequence of data items).

We can now understand an important distinction between traditional RPC and gRPC: the interactions they support. A traditional RPC follows a request-response interaction: each call to a remote procedure causes a single message to travel over the network to the computer containing the remote procedure and a single message to travel back. gRPC extends remote procedure call to allow a remote procedure to stream multiple data items in response to a request.

Microservices have also used variations of basic communication interactions. For example, some microservices follow the *publish-subscribe* variant of data streaming†. An entity contacts a microservice and specifies a topic. The network connection remains open perpetually, and the microservice sends any data items that arrive for the specified topic. Other entities contact the microservice to send data items; each data item is labeled with a topic. The name arises because the senders are said to "publish" information on various topics, and receivers are said to "subscribe" to specific topics.

12.10 Using A Service Mesh Proxy

When creating a microservice, a designer does not need to use a single technology or a single interaction for all communication. Instead, a separate choice can be made for each aspect of communication. For example, a microservice that uses gRPC to communicate with other microservices may choose to expose a REST interaction for entities that use the microservice. In addition, recall that the microservice approach allows each microservice to be scaled independently. That is, additional instances of a microservice can be created as needed. Several questions arise:

- What software creates instances of a microservice?
- How are requests forwarded to a given instance?
- How can requests be translated to an internal form?
- How does one microservice discover another?

Industry uses the term *service mesh* to refer to software systems that handle such tasks. One aspect stands out: forwarding of requests. A service mesh uses a straightforward approach: instead of allowing entities to contact an instance of a microservice directly, have the entities contact a *proxy* that manages a set of instances and forwards each request to one instance. Figure 12.7 illustrates a service mesh proxy.

---

†For example, the *Object Management Group* (*OMG*) defines a *Data Distribution Service* (*DDS*) that uses publish-subscribe.

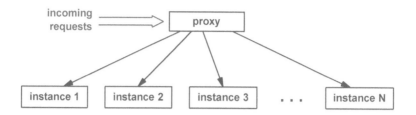

Figure 12.7 Illustration of a proxy for a service mesh. To use the service, an external entity contacts the proxy.

Note that the use of a service mesh proxy answers the question about differences between the external and internal communication interactions and technologies. In the figure, for example, the proxy might expose a REST API to entities that use a microservice, and might use gRPC when communicating with other microservices. The point is:

The use of a proxy allows a microservice to be scaled and isolates the communication used internally from the communication used to access the microservice.

12.11 The Potential For Deadlock

Although the microservice approach has many advantages, a distributed system composed of many microservices can fail in unexpected ways. Such designs are especially susceptible to circular dependencies where a set of microservices all depend on one another. If each microservice in the cycle is waiting for another microservice, a *deadlock* can result.

It may seem that automated scale out makes microservices immune to deadlock. Suppose microservice *A* depends on microservice *B*. Even if an instance of *B* remains blocked, when *A* attempts to use *B*, another instance of *B* will be created. With thousands of independent microservices, however, circular dependencies can arise in subtle ways that may not be obvious to the software engineers who create and maintain the microservices.

To see how circular dependencies can be introduced, consider a trivial example of four microservices: a time service, a location service, a file storage service, and an authentication service. Although the services access one another, the system runs with no problems. The file storage service uses the time service to obtain the time of day that it uses for timestamps on files. The location service uses the file storage service to obtain a list of locations. The authentication service uses the location service to find the location of the file storage service (which it uses to obtain stored encryption keys). Interestingly, if any service is completely terminated and restarted, the system continues to work correctly as soon as the restart occurs.

Suppose that the software engineer in charge of the time service decides to authenticate each request. That is, when it receives a request, the time service will contact the authentication service and wait for approval before answering the request. Once the time service is restarted running the new code, the system continues to operate correctly. However, a subtle circular dependency has been created in the boot sequence that can cause a deadlock. Figure 12.8 illustrates the dependency cycle and a series of events that can cause a deadlock.

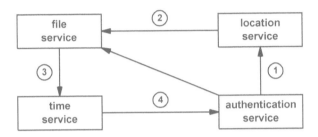

1. The authentication requests file service location
2. The location service requests its initial configuration
3. The file service requests the time of day
4. The time service tries to authenticate access

Figure 12.8 A dependency cycle among four microservices.

Interestingly, a deadlock will not occur unless all four microservices attempt to start at the same time. Normally, only one microservice will restart at any time, and the system will continue to operate. However, if a massive power failure occurs, all services will restart simultaneously, and a deadlock can result.

Note that because each microservice is designed and implemented independently, a software engineer who modifies one service may be unaware of how the modification will affect dependencies. Moreover, the effect can remain hidden until an unusual situation arises, such as restarting after a power failure. The important point is:

Because microservices are developed and maintained independently, circular dependencies can remain hidden.

12.12 Microservices Technologies

Many technologies have been created to aid software engineers in the design and operation of microservices. For example, commercial and open source service mesh technologies exist, including *Linkerd*, a project of the *Cloud Native Computing Foundation* (*CNCF*), and *Istio*, a project joint among Google, IBM, and Lyft. In addition, many frameworks exist that help developers create and manage microservices, including *Spring Boot*.

12.13 Summary

The microservices approach disaggregates a monolithic application into multiple services, allowing them to scale independently. The approach has advantages for software development, including smaller scope, smaller teams, less complexity, a choice of programming language, and more extensive testing. The approach has advantages for operations and maintenance, including rapid deployment, improved fault isolation, better control of scaling, compatibility with containers and orchestration systems, and independent upgrade of each microservice.

The approach also has potential disadvantages, including cascading errors, duplication of functionality and overlap, management complexity, replication of data and transmission overhead, an increased security attack surface, and the need for workforce training.

Microservices communicate over a network using transport protocols, such as HTTP and gRPC. HTTP supports a request-response (REST) interaction. gRPC generalizes conventional RPC to provide a streaming interface in addition to remote procedure invocation and return.

Service mesh software automates various aspects of running a microservice. The use of a proxy for the service hides both the internal structure and internal communication protocols, which allows the protocols used for external access and internal microservice invocation to differ.

An important weakness of the microservices approach arises because each microservice is created and maintained independently. Subtle problems, such as circular dependencies, can arise and remain hidden until an unusual event occurs, such as a power failure that causes all microservices to restart simultaneously.

Chapter Contents

13   Controller-Based Management Software

13

Controller-Based
Management Software

13.1 Introduction

Previous chapters describe paradigms used to create cloud-native software systems, including the MapReduce and microservices paradigms. This chapter focuses on software that automates the management of resources. The chapter explains controller-based designs and concepts of declarative specification and automated state transition.

13.2 Traditional Distributed Application Management

An understanding of controller-based designs begins with understanding how traditional distributed systems are managed. For example, consider a distributed application deployed throughout an organization. Suppose the application maintains information about employees. One way to deploy such an application requires each department to run an instance of the application that stores data about the employees in the department†. Authorized users run client applications that access the instance. The design works especially well if most accesses have high locality (i.e., accesses that originate in a given department usually refer to employees in the same department). Using multiple instances means the system is general because it allows an authorized user in any department to access information about an employee in other departments, and the system is efficient because most accesses go to the local copy.

How does an IT staff manage such a distributed system to ensure that all instances remain available? A popular approach uses a human operator and an automated monitoring tool. The idea is straightforward: the tool continually monitors each of the instances and alerts the human operator if a server stops responding. The operator can

---

†Technically, each instance runs as a *server*.

then investigate the problem and take action. If an instance has crashed, the operator can reboot the server. If a network has failed and communication has been cut off, the operator can diagnose and repair the problem. The point is:

> *Because a traditional distributed system runs instances on many physical computers, managing such a system usually employs a monitoring tool that alerts a human operator when a server application stops responding.*

13.3 Periodic Monitoring

How does an automated monitoring tool work? To test an instance, many tools send a special management message to which the application responds. If no response arrives, the tool might retry once and then alert the operator about the problem. To avoid flooding an instance with a continual stream of monitoring requests, most tools check periodically. For example, a tool might check each instance once every thirty seconds.

A monitoring tool runs in background and never terminates. The code is arranged to repeat a set of steps indefinitely. Algorithm 13.1 lists the steps taken to monitor the instances in a traditional distributed system.

Algorithm 13.1

Purpose: Monitor a distributed system

Given: A set of application instances for the system

Action: Alert an operator if any instance fails

Method: Check each instance every thirty seconds

Steps:

    Repeat the following every 30 seconds {

        Send a message to each application instance

        If an instance fails to respond, send an

            alert to the operator

    }

Algorithm 13.1 A tool that monitors a traditional distributed system.

Some management systems provide support for periodic execution of a monitoring application. In such systems, the algorithm can be implemented by writing a piece of code for the two steps and configuring the system to execute the code once every 30 seconds. In other systems, the algorithm is implemented as a background process. To prevent the process from terminating, the code must specify an infinite iteration informally known as an *infinite loop*. To perform an action periodically rather than continually, the designer must insert an explicit delay in the loop†. For example, Figure 13.1 shows the arrangement of code used to implement Algorithm 13.1 as a process.

```
loop forever {     /* known as an infinite loop */
            check each of the servers
            if a server fails to respond, send
                    an alert to the operator
            delay for 30 seconds

}
```

Figure 13.1 The implementation of a periodic monitoring tool as a process with an explicit delay used to limit execution to once every 30 seconds.

13.4 Managing Cloud-Native Applications

Most cloud-native applications use container replication to scale out services. Can a system similar to Algorithm 13.1 be used to monitor cloud-native applications? Unfortunately, a cloud-native application is much more complex than a traditional distributed system. To understand the complexity, consider a few differences.

First, consider the difference in instances. A traditional distributed application consists of a fixed set of static instances. For example, the employee information system described above has one instance for each department in an organization. By contrast, cloud-native applications deploy instances dynamically as needed, creating additional instances to scale out an application. During times of low demand, only a few instances remain active, and during times of high demand, many instances remain active. Thus, unlike a tool that monitors a traditional distributed system, a tool that monitors a cloud-native application must avoid sending erroneous alerts about instances that have been shut down to decrease the scale.

Second, consider the difference in instance locations. In a traditional distributed system, the location of each instance is known in advance and never changes. In the employee information system, for example, each department runs its copy of the application on the same computer in its department. For a cloud-native application, however, orchestration software uses the current load on each server and other factors to choose where to deploy an instance.

---

†A later section in the chapter explains a related concept that also uses the term *loop*.

Third, consider the structure of applications. Unlike traditional distributed systems in which each application is constructed as a monolithic program, cloud-native applications consist of disaggregated programs that run as multiple microservices. Thus, instead of monitoring multiple copies of a single application, software built to manage a cloud-native application must monitor multiple instances of multiple microservices.

Fourth, consider application persistence. A traditional application instance executes as a server process that starts when a computer boots, and remains running until the computer shuts down. In contrast, cloud-native software runs in containers. Typically, a container is designed to service one request and then terminate; a new container is created for each request, possibly at a new location. Thus, monitoring a cloud-native service requires handling extremely rapid changes in the set of active instances.

Figure 13.2 summarizes differences between a traditional distributed application and a cloud-native application that make managing a cloud-native application complex.

Traditional Application	Cloud-Native Application
The number of instances does not change dynamically	The number of instances changes as an application scales out
The location of each instance is known in advance	The locations of instances are chosen dynamically to balance load
Each instance consists of a monolithic application	An application is disaggregated into multiple microservices
An instance runs as a process that persists indefinitely	An instance runs as a container that exits after handling one request

Figure 13.2 A comparison of traditional and cloud-native applications.

The point is:

Monitoring a cloud-native application involves monitoring multiple microservices that are each composed of a varying number of instances at changing locations, with each instance only persisting for a short time.

13.5 Control Loop Concept

The term *control loop* originated in control theory, and is used in automation systems to refer to a non-terminating cycle that adjusts a system to reach a specified state. For example, consider a residential HVAC system with a thermostat used to regulate temperature. A user specifies a temperature, and the HVAC system automatically adjusts the temperature to match the value the user specified. In terms of a control loop, the user specifies a desired state, and the control loop in the thermostat adjusts the system to make the actual temperature match the value the user selected. Thus, a control loop implements a declarative, intent-based interface in which the user merely specifies the intended result and the control loop implements the changes needed to achieve the result.

We can imagine the thermostat repeating three basic steps: measurement, comparison, and performing heating or cooling. The cycle starts by measuring the current temperature. The thermostat then compares the measured value to the temperature the user has specified. If the two temperatures differ, the thermostat instructs the system to heat or cool the air to move the system toward the desired temperature; if the two temperatures are equal, the system has reached the desired state, and the thermostat instructs the system to stop heating or cooling.

We use the term *control loop* because the steps taken form a conceptual cycle. Figure 13.3 illustrates the idea, and shows how the steps can be arranged in a loop,

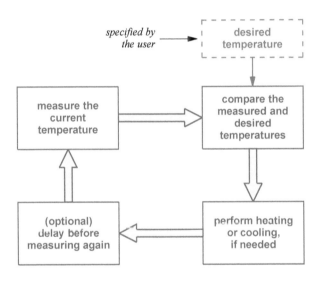

Figure 13.3 The conceptual cycle of a control loop for a thermostat that regulates temperature.

13.6 Control Loop Delay, Hysteresis, And Instability

The delay step in a control loop is optional, but can be important for two reasons. First, taking measurements too rapidly can waste resources and lead to unexpected behavior. For example, consider a remote wireless thermometer that runs on battery power. Because temperature changes slowly, transmitting measurements continually without delay wastes battery power.

As an example of unexpected behavior, consider a control loop that positions a robot arm. Suppose a measurement finds that the arm is currently X millimeters to the left of the desired position. The control loop will send a command that moves the arm X millimeters to the right. Moving a mechanical arm takes time. If the control system measures the arm position immediately after sending the command, the arm will only have started to move, and will remain Y millimeters to the left of the desired position. Therefore, the control loop mechanism will send an additional command to move the arm Y millimeters to the right. As a result, the arm will travel X + Y millimeters total, leaving it too far to the right. If the control includes a delay step that pauses long enough for a motion command to take effect, the next measurement will represent an accurate position, avoiding unnecessary motion.

We use the term *hysteresis* to refer to the lag between the time a change is initiated and the time it takes effect†. Hysteresis is important because it can cause unexpected results. The robot arm example shows that a second iteration of the control loop can send an unwanted command to move the arm. To understand the worst case, consider a control loop that measures the position ten times a second, and suppose it takes two seconds to move the arm. The control loop will make twenty measurements while the arm is moving once, and will send a command for each. If the underlying hardware buffers commands, each of the commands will eventually be honored, and the arm will move much too far.

Interestingly, once it has detected that the robot arm has moved too far, the control loop will attempt to correct the problem by sending commands that move the arm back in the opposite direction. Depending on the speed of the loop, overcompensation can occur for the reverse motion —delay in moving the arm can result in multiple commands being sent, and the arm will eventually move too far in the reverse direction. Thus, the arm can oscillate back and forth, never stopping at the desired position.

In a badly designed system, instead of converging on the desired state, oscillations can cause a system to run wild. For the robot arm, instead of moving toward the desired position, the distance can increase, causing the arm to move farther and farther away from the desired position on each oscillation. The point is:

> *If a control loop takes measurements before a change has time to take effect, the results can be unexpected; adding a delay to the loop may be necessary to prevent oscillations and guarantee that the system convergences on the desired state.*

13.7 The Kubernetes Controller Paradigm And Control Loop

A programmer who writes traditional control loop code for an IoT device must plan the loop carefully to prevent unintended behavior that can result from taking measurements too quickly. Interestingly, cloud orchestration systems, including Kubernetes, employ a variant of a control loop that eliminates periodic measurements. This section describes the variant, and Chapter 16 further explains its use with some IoT devices.

---

†Our use is informal; the term comes from physics where it refers to the lag between the time a magnetic field changes around a magnetic material (such as iron) and the time the magnetism of the material changes.

Recall from Chapter 10 that Kubernetes uses a set of *controllers* to automate various management tasks, such as scale out. We can now understand how a controller can manage container deployments and microservices: the software employs an intent-based control loop. We use the term *controller pattern* to characterize the design. Conceptually, each Kubernetes controller:

- Runs a control loop indefinitely
- Compares the actual and desired states of the system
- Makes adjustments to achieve the desired state

Algorithm 13.2 captures the essence of a Kubernetes controller by showing how it follows a declarative, intent-based paradigm to manage a microservice.

Algorithm 13.2

Purpose: Act as a controller that manages a microservice
consisting of multiple pods of containers

Given: A specification of a desired state of the service

Method: Continually obtain information about the actual
state of the service. Compare the actual state to
the desired state, and make adjustments to move
the service toward the desired state

Algorithm 13.2 A Kubernetes controller for a microservice.

13.8 An Event-Driven Implementation Of A Control Loop

How can Kubernetes obtain accurate information about the actual state of the system without repeatedly polling to take measurements? The answer lies in an *event-driven* control loop. Instead of arranging for the controller to check status periodically, an event-driven control loop configures components to inform the controller whenever a change occurs.

Typically, event-driven systems use message passing. When a change occurs, the controller receives a message about the change. For example, when the specification file changes, a file system component sends the controller a message. Similarly, the orchestration component sends the controller a message when an instance exits.

In essence, using an event-driven approach reverses the responsibility for obtaining information about the state of the system. In a traditional control loop, the controller actively polls the system to obtain the information; in an event-driven control loop, the controller waits passively to be informed when a change occurs. Figure 13.4 shows the steps an event-driven controller follows to react to incoming messages.

Read and parse the specification file
Use the specification to create pods for the system
loop forever {
    wait to receive a message
    if the specification changed, read and parse the file
    If the status of a pod has changed, record the change
    if the change leaves the service outside the desired state,
        adjust the system to move toward the desired state
}

        Figure 13.4 An event-driven implementation of a Kubernetes controller in which the controller waits passively to be informed when a change occurs.

Note that using the event-driven paradigm means that a controller does not perform a delay step as part of the control loop. From a programmer's point of view, the approach means that there is no need to choose a delay, and there is never a problem with a delay that is too small or too large. Furthermore, unlike a traditional event loop that continues to generate processing and communication overhead even when no changes are occurring, an event-driven design does not consume resources unnecessarily. We can summarize:

> *The controller paradigm used by Kubernetes follows a declarative, intent-based approach in which a user specifies the desired state of a service and a controller continually adjusts the system to move toward the desired state. Using an event-driven implementation for a controller avoids needless polling.*

13.9 Components Of A Kubernetes Controller

Conceptually, Kubernetes runs multiple, independent controllers. In practice, Kubernetes implements all controllers with a single background daemon process, the *kube-controller-manager*. Kubernetes divides each controller into three pieces†:

- Informer or SharedInformer (watcher)
- Workqueue and Dispatcher
- Workers

---

†See Figure 14.4 on page 201 for an illustration of the conceptual organization of the pieces.

Informer or SharedInformer. An Informer, also called a *watcher*, looks for changes in the state of Kubernetes objects and sends an event to the Workqueue whenever a change occurs. Although it is possible to use a traditional control loop that periodically requests state information from the Kubernetes API server, the use of informers reduces overhead. To avoid polling a list to find new items, Kubernetes includes a *Listwatcher* component that generates a notification about the creation of a new instance or a change in a current instance.

An Informer component keeps a local cache of the state of resources. Many microservices use a set of controllers to handle multiple aspects of the service. In such cases, *SharedInformer* provides an alternative that allows cached information about the state of the service to be shared among a set of controllers.

Workqueue and Dispatcher. The Workqueue contains a list of changes that have occurred in the state of the service. New items are added when a change occurs, and a *Dispatcher* removes items as they are processed. When an item appears on the Workqueue, it may mean that the current state of the service no longer adheres to the desired state.

Workers. When it extracts an item from the Workqueue, the Dispatcher checks the specification. If the system no longer conforms to the desired state, the Dispatcher invokes a worker to make adjustments to move toward the desired state (e.g., by creating a new instance to replace one that has completed). As with most cloud facilities, it is possible to scale a controller by creating multiple workers.

How can Kubernetes know the steps that should be taken to align the state of a computation with the desired state? It cannot. Instead, a programmer creates code that performs the necessary actions. The programmer writes a function, *Reconcile*, that Kubernetes calls. When it runs, Reconcile will have access to the specification, and can adjust the state of the system accordingly.

Although the description above implies that the controller operates continuously, Kubernetes does provide a way for a controller to run periodically. A value known as *ResyncPeriod* specifies when a controller should revalidate all items in the cache. In essence, ResyncPeriod specifies the time at which the controller compares the current state to the desired state and takes action accordingly. Of course, a designer must be careful not to set ResyncPeriod too low or the controller will incur a high computational load needlessly.

13.10 Custom Resources And Custom Controllers

Kubernetes defines a set of core resources and controllers that cover most of the common tasks associated with deploying and managing a cluster that includes multiple pods. Despite the broad scope of the built-in facilities, a user may wish to define a new resource or a new controller to handle a special-purpose task, possibly interacting with systems outside of Kubernetes. Therefore, Kubernetes provides mechanisms that allow a user to define custom resources and build custom controllers that extend the built-in facilities. A customized controller can manage built-in resources or custom resources;

the combination of a new custom resource and a custom controller to manage the resource can be useful.

Like built-in controllers, a custom controller employs the event-driven approach. To create a custom controller, a software engineer must write code for the basic components: a Workqueue, Worker(s) to process events, and either a SharedInformer, if the controller maintains information about multiple pods, or an Informer/watcher, if the controller has small scope. Although creating such components may seem complex, the kube-controller-manager handles many of the details. Thus, a software engineer can focus on the code related to the application, and does not need to write code to store events or for the control loop. To further make the task easy, many examples and tutorials exist that explain the steps needed. Interestingly, a custom controller can run as a set of pods or as a process that is external to Kubernetes. The point is:

> *In addition to an extensive set of built-in controllers and pre-defined resources, Kubernetes allows a user to define custom resources and custom controllers.*

13.11 Kubernetes Custom Resource Definition (CRD)

Kubernetes provides a facility that helps a software engineer create a custom resource. Known as the *Custom Resource Definition* (*CRD*), the facility allows one to define new objects and then integrate them with a Kubernetes cluster. Once it has been defined, a new object can be used exactly like a native Kubernetes object.

To understand how CRD can help raise the level of abstraction, consider an example. Suppose an application uses a database. A variety of database technologies exists, and the details differ. Although it is possible to build an application for one specific technology, doing so limits generality and means the application must be rewritten before it can be used with a different type of database.

Various approaches have been used that allow an application to access multiple underlying facilities. For example, Figure 13.5 illustrates a *proxy service*.

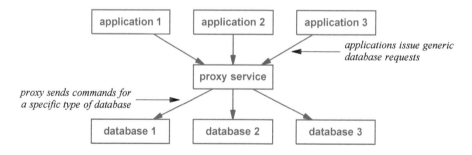

Figure 13.5 Illustration of a proxy service that allows applications to access multiple types of databases.

As the figure shows, a proxy service fits between applications and underlying databases. Instead of accessing a database directly, an application invokes the proxy service. Each database technology defines a specific set of commands that must be used to communicate with the database (i.e., a database-specific API). The proxy accepts a generic set of database requests from applications. To fulfill a request, the proxy translates the request into database-specific commands, and issues the commands to the underlying database. That is, a proxy service offers a single, generic interface that applications use, and only the proxy needs to understand the details of underlying databases. The key point is that an application can switch from one database to another without being modified.

The CRD facility in Kubernetes offers the same benefits as a proxy without requiring a user to create and manage a separate service. To use CRD, a software engineer creates a Custom Resource Definition that accepts generic database commands just as a proxy does. The CRD might be named *Database*. The software engineer creates a specification for the CRD along with code that translates generic requests into database-specific commands (e.g., for *Redis* or *MongoDB*). Thus, a Database CRD functions like a proxy service, but instead of running and managing a separate service, Kubernetes can manage instances of the CRD.

13.12 Service Mesh Management Tools

Despite all the facilities in Kubernetes, additional tools are available that offer ways to manage services. Management functions include service discovery, health checking, secure communication among a set of services, key-value object storage, and support for deploying an application across multiple data centers. Examples include *HashiCorp Consul*, *Istio*, and *Linkerd* from the Cloud Native Computing Foundation (CNCF) project.

Most service mesh tools have been designed to work with Kubernetes. One of the main arguments in favor of mesh management tools arises from the need for security. Kubernetes provides a way to deploy services, and a mesh management tool can ensure that communication among services remains secure. In addition, a mesh management tool may offer additional control functions for Kubernetes clusters.

13.13 Reactive Or Dynamic Planning

The demands on cloud applications can change quickly, and we use the term *reactive planning* to refer to the rapid planning required to accommodate new conditions. Because the term *reactive* has a somewhat negative connotation compared to *proactive*, some managers say that cloud management systems conduct *dynamic planning*.

The idea of reactive/dynamic planning is straightforward: adapt quickly to the constant stream of changes in both the environment and demand by planning a new desired state and moving the service to the desired state. The controller paradigm and control loops can be used to implement reactive/dynamic planning.

Interestingly, constant change may mean that a particular service never reaches a steady state. Instead, the management system adapts to continual changes by constantly revising the desired state of the service. Provided controllers are designed to make useful decisions and accommodate any hysteresis introduced by queues, the service can continue to change without becoming unstable.

13.14 A Goal: The Operator Pattern

Most controllers handle routine, repetitive tasks. Industry uses the term *operator pattern* to describe an envisioned control system that can handle the remaining management tasks that normally require a human operator, such as identifying anomalies, diagnosing problems, understanding how and when to apply and deploy software updates, and how to delete resources associated with a service when the service is shut down. Software that achieves the envisioned goal will require Artificial Intelligence, and the idea of using the operator pattern to build a control system is related to the goal of using AIops mentioned in Chapter 9†.

13.15 Summary

Managing a cloud-native application is inherently more complex than managing a traditional distributed system. In a cloud-native application, the number and location of instances changes, a single application may be disaggregated into multiple microservices, and instead of persisting, an instance exits after handling one request.

Borrowing the concept and terminology from control theory, we use the term *control loop* to refer to a non-terminating, intent-based computation that continually measures the state of a system, compares the actual state to the desired state, and makes adjustments to move the system toward the desired state.

Kubernetes provides a set of built-in controllers that each run a control loop for one particular aspect of a cluster. A Kubernetes controller consists of three components that implement an event-driven approach: an Informer/watcher (or SharedInformer) that sends events when the state of the system changes, a Workqueue that holds a list of events, and a Worker that handles events.

Kubernetes offers users the ability to define custom resources and custom controllers. A Custom Resource Definition (CRD) can be used to create a generic interface for applications and then allow each instance to use a specific underlying technology.

---

†See page 119 for a definition of AIops.

Chapter Contents

14

Serverless Computing And Event Processing

14.1 Introduction

Previous chapters describe algorithms, platforms, and technologies that can be used to create cloud-native software systems. This chapter considers facilities cloud providers offer that enable programmers to create, deploy, and scale applications quickly, easily, and at low cost. The chapter explains why such facilities have gained popularity and can reduce costs for a cloud customer that offers a service.

14.2 Traditional Client-Server Architecture

To understand the advantages of serverless computing, one must first understand the traditional client-server architecture that network applications use. We will review the basic ideas and then consider the expertise required to deploy and operate a server.

When application programs communicate over a network, they follow the *client-server paradigm*, which divides applications into two categories based on how they initiate communication. One of the two programs must be designed to start first and remain ready to be contacted. The other application must be designed to initiate contact. We use the terms *client* and *server*† to describe the two categories:

- Server: an application that runs first and waits for contact
- Client: an application the contacts a server

---

†Unfortunately, ambiguity arises because data centers use the term *server* to refer to a computer powerful enough to run a server application.

Note: the categories only refer to the initial contact between the two: once network communication has been established, data can flow in either direction.

In the simplest case, two computers attach to a network or to the Internet. A client application running on computer 1 contacts a server application running on computer 2, and the two communicate. Figure 14.1 illustrates the idea.

Figure 14.1 An example of traditional client-server communication.

14.3 Scaling A Traditional Server To Handle Multiple Clients

In a traditional client-server architecture, a server scales by using concurrent execution to handle multiple clients at the same time. That is, the server does not handle one client and then go on to the next. Instead, the server repeatedly waits for a client to initiate communication and then creates a concurrent process to handle the client. Thus, at any time, $N+1$ server processes are running: a *master* server that clients contact, and N instances of the server that are each handling an active client. To handle multiple clients, the computer on which a server runs must have more memory and processing power than a computer that runs a client. Figure 14.2 illustrates a concurrent server handling three clients.

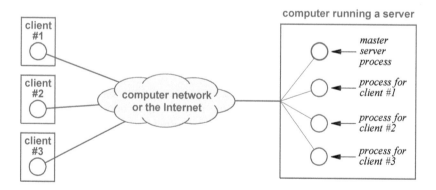

Figure 14.2 An example of traditional client-server communication.

The figure makes it appear that four independent application programs are running on the server computer. In fact, they are all copies of one server application. When a connection arrives from a new client, the master server copy calls *fork* (or the equivalent) to create a new process to handle the client.

Conceptually, the application for a concurrent server integrates two aspects of a server into a single application program:

- Fulfilling the service being offered
- Replicating and scaling the server

Fulfilling the service. A client contacts a server to access a service. Typically, a client sends a request to which the server sends a reply. The chief function of a server lies in interacting with clients to fulfill clients' requests.

Replicating and scaling. The second aspect of a server arises from the techniques used to replicate and scale the server. A traditional concurrent server uses contact from a new client to trigger the creation of a separate process to handle the client. The important point is:

> *In a traditional server, a single application contains code to replicate the server to handle multiple simultaneous clients as well as code to interact with a given client and handle the client's requests.*

14.4 Scaling A Server In A Cloud Environment

A traditional concurrent server replicates copies of itself automatically to handle multiple clients simultaneously. However, all copies must run on the same physical computer. The approach does not work well in a cloud environment because cloud systems achieve large scale by replicating instances across many physical machines. Consequently, complex software systems must be used to run servers in a cloud data center. The systems handle instance management by deploying copies as needed and must also handle network communication by arranging to forward traffic from each client through a proxy to the correct instance. Figure 14.3 illustrates management software controlling a deployment.

14.5 The Economics Of Servers In The Cloud

As previous chapters describe, a large set of management technologies exist that can be used to deploy and operate services, including orchestration systems, proxies, load balancers, and service mesh management software. In terms of cost, many of the software technologies follow the open source model, making them free. Thus, it may seem that open source software allows a cloud customer to port existing servers to the cloud and then deploy and scale the servers at little extra cost. That is, a customer only needs to have basic server software, lease a set of VMs, and use open source software to handle deployment and scaling.

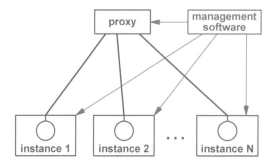

Figure 14.3 Management software controlling a proxy and replicas of a server on multiple physical machines.

Unfortunately, two costs can be significant:

- Unused capacity
- Expertise and training

Unused capacity. Because setting up a VM takes time, a customer must plan ahead by allocating sufficient VMs to handle the expected peak load. As a result, the customer must pay for VMs that remain idle during off-peak times.

Expertise and training. Although open source management systems are free, using such technologies effectively and safely requires a significant amount of expertise that lies outside the typical expertise of a software engineer. Thus, a customer must hire experts. A customer that uses multi-cloud, must have expertise for each cloud system. Furthermore, because cloud technologies continue to evolve, the customer must pay for training to keep their staff up to date.

14.6 The Serverless Computing Approach

In response to the need for server management, cloud providers created a way that a cloud customer can pay to have the provider handle all tasks associated with running the customer's servers. The industry has adopted the name *serverless computing* to refer to the approach. To an outsider, the name seems inappropriate because the underlying system does indeed run servers and uses the client-server model of interaction. However, providers assert that from a customer's point of view, the facilities are "serverless" in the sense that the customer can build and run software to fulfill users' requests without thinking about the deployment and replication of servers, without configuring network names and addresses for servers, and without leasing VMs to run the servers. To help eliminate confusion, the industry sometimes uses the alternative name *Function as a Service (FaaS)*†. The name arises because a cloud customer writes code that performs the main function of the server and allows the provider to handle deployment. To summarize:

---

†The terms *Backend as a Service (BaaS)* and *Runtime as a Service (RaaS)* have also been used to specify that the provider handles mechanisms needed to run a server.

Also known as Function as a Service *(FaaS), the serverless computing approach allows a cloud customer to avoid dealing with servers by paying a cloud provider to deploy and scale the customer's servers.*

Why would a customer pay a provider to deploy servers? The customer can reduce overall cost. First, the customer can avoid charges for unused capacity because the provider only charges fees for the time a server is being used. That is, instead of creating VMs, the provider only runs a copy of a server when needed. Second, a provider amortizes the cost of experts and their training over all customers, making it less expensive for each customer. The reduction in local expertise can be especially significant for customers that follow the multi-cloud approach of using more than one provider. The point is:

Because a provider amortizes the cost of expertise among all customers and only charges for the time a server is used, the serverless approach can cost a customer less than running servers themselves.

What happens if no one uses a given server? In some cases, a provider imposes a minimum monthly fee. However, other providers offer service without a minimum fee, meaning that if no one uses a server in a given period, the customer pays nothing. Industry uses the term *scale to zero* to refer to such services.

In terms of accommodating large numbers of clients, the serverless approach offers great flexibility. Unlike a self-managed set of servers that require a customer to plan for a peak load, serverless computing allows arbitrary scale. Industry uses the term *scale to infinity* to characterize a serverless system that does not place a limit on scale. In practice, of course, a provider must set some limits, but the idea is that the limits lie beyond what a typical customer will ever need.

14.7 Stateless Servers And Containers

How can a provider offer serverless computing that scales from zero to infinity and achieves both efficiency and low cost? The answer lies in combining technologies and architectures covered in previous chapters. To make it feasible to deploy servers quickly, serverless technologies use containers (Chapter 6). To deploy and manage a server, serverless systems use orchestration (Chapter 10), and to handle scale out, a serverless system uses the controller-based approach (Chapter 13).

Despite building on extant technologies, the serverless approach introduces two key features that distinguish it from the traditional server approach:

- The use of stateless servers
- Adherence to an event-driven paradigm

The use of stateless servers. We use the term *state* to refer to data related to clients that a server stores internally. The term *stateful server* refers to a server that stores state information; and the term *stateless server* refers to a server that does not store state information. Stateful servers store information for two reasons. First, keeping information related to a given client allows a server to provide continuity across multiple contacts by the client. Second, keeping state information allows a server to share information among multiple clients. A stateful approach works well for a traditional server because the server runs on a single computer. Therefore, a traditional server can use mechanisms such as shared memory to allow all instances of the server to access and update the state information. In particular, a design that uses threads allows all instances to share global variables. Furthermore, because a traditional server has a long lifetime, state information usually persists across many client contacts.

The stateful approach does not work well for a server that runs in a data center and handles large scale. To achieve scale, orchestration systems deploy instances on multiple physical computers, making shared memory impossible. To handle microservices, containers are designed with a short lifetime: the container starts, performs one function, and exits. Thus, state information does not persist for more than one client connection. To capture the idea that serverless computing focuses on running a single, stateless function in each container (i.e., FaaS), some engineers say that serverless computing runs *stateless functions*.

> *Because it uses containers and can run on multiple physical servers, a serverless computing system requires server code to be stateless.*

It is important to distinguish between a stateful server and a server that keeps data in a database or persistent storage. Statefulness only refers to the information a server keeps in memory while the server runs. When a server exits, state information disappears. In contrast, data stored on an external storage system (e.g., a database, a file on NAS, or an object store) persists after the server exits; a later section contains an example. A server can use external storage to achieve the same effect as statefulness —the server saves a copy of internal data before exiting and reads back the data when it starts running. Stored data does not count as state information because the data is not lost when the server exits. The point is:

> *Although serverless computing requires servers to follow a stateless design, a server may store and retrieve data from a database or persistent storage, such as a file on NAS or an object store.*

Adherence to an event-driven paradigm. The previous chapter describes controller-based management software in general and the event-driven paradigm that Kubernetes controllers use. Serverless computing adopts the paradigm, and generalizes it. The underlying cloud system generates events when changes occur (e.g., when a physical server fails). In addition, serverless systems count each server access as an event. For example, in addition to a REST interface that accepts contact via HTTP, some serverless systems provide *management interface* or other *programmatic interface* components that allow a human user to interact with the system or a computer program to use a protocol other than HTTP. A contact from any of the interfaces counts as an event.

14.8 The Architecture Of A Serverless Infrastructure

Serverless computing adopts the technology Kubernetes uses for controllers and follows the same general architecture. The chief components include an event queue, a set of interface components that insert events into the queue, and a dispatcher that repeatedly extracts an event and assigns a worker node to process the event. In the case of serverless computing, worker nodes run the server code. Figure 14.4 illustrates the components in a serverless intrastructure.

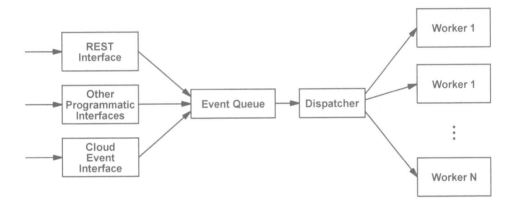

Figure 14.4 The architecture that providers use for serverless computing.

14.9 An Example Of Serverless Processing

Netflix uses the AWS *Lambda* event-driven facility for *video transcoding*, a step taken to prepare each new video for customers to download†. The arrangement has become a canonical example of serverless computing. Figure 14.5 lists the basic steps taken to transcode a video.

---

†See *https://aws.amazon.com/solutions/case-studies/netflix-and-aws-lambda/* for further explanation.

Step	Action Taken
1.	A content provider uploads a new video
2.	A serverless function divides the new video into 5-minute segments
3.	Each segment is given to a separate serverless function for processing
4.	The processed segments are collected and the video is available for customers to access

Figure 14.5 The basic steps Netflix uses to transcode a video.

Events trigger each of the serverless processing steps. When a content provider uploads a new video, the system places the new video in an Amazon S3 bucket. The S3 object storage system generates an event that triggers a serverless function to divide the video into segments that are each five minutes long. When a segment arrives in an S3 bucket, another event triggers a serverless function that processes and transcodes the segment. Thus, transcoding can proceed in parallel. Figure 14.6 illustrates how video data flows through the system and how events trigger processing.

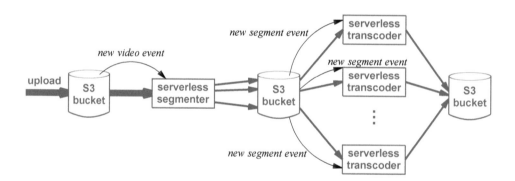

Figure 14.6 The Netflix transcoding system. A new video event causes the video to be divided into segments; a new segment event causes the segment to be transcoded.

14.10 Potential Disadvantages Of Serverless Computing

It may seem that compared to traditional server designs, serverless computing offers three unbeatable advantages: the ability to scale arbitrarily, no need to manage servers, and lower overall cost. Despite its advantages, serverless computing does have potential disadvantages. First, serverless systems introduce latency. A traditional server always starts before any client initiates contact. Thus, the server can respond to requests immediately. In a serverless situation, however, management software launches instances of the server as needed. The delay in starting a server may seem insignificant, but can be important for a server that needs to respond quickly (e.g., to handle an event or a transaction).

Surprisingly, serverless computing can also generate unexpected costs. To see why, consider serverless computing used with a disaggregated microservice. The serverless approach makes it easy to divide processing into small pieces that each run independently. In fact, the concept of Function as a Service can lure software engineers into running each function as a separate piece. Because a provider charges for each instance, however, dividing a program into tiny pieces means the user will incur a small charge for each piece. For example, suppose a program contains four functions A, B, C, and D and for typical input A calls B, B calls C, and C calls D:

$$A \rightarrow B \rightarrow C \rightarrow D$$

If each function runs as a separate entity, a customer will be charged to create four instances whenever A is invoked. Placing all four in the same piece means a customer will only pay for one instance. Although the charge for an instance is trivial, a microservice architecture allows a given microservice to be invoked thousands of times per second, in which case a factor of four in cost can accumulate quickly.

A related disadvantage of the serverless paradigm can arise when a software system is designed to process events. Consider, for example, alerts sent to a human operator to report performance problems. Each program that uses a service sends an alert if the service is unavailable. Suppose a program uses microservice X, microservice X uses Y, and Y uses Z. Now suppose Z has failed. Y will send an alert to the operator saying that Z has failed and will respond to X that it cannot fulfill the request. X will send an alert to the operator saying that Y failed and will report failure to its caller. The point is that a failure at a low level can result in a cascade of alerts. Because microservices can be invoked thousands of times per second, the system may receive an unexpected flood of events. If the alert system employs the serverless approach, the system will scale out by creating an instance for each alert. Thus, the charges incurred to scale the system can become surprisingly high.

We can conclude:

Despite its many advantages, the serverless paradigm has disadvantages of introducing a small delay when starting a server and the potential for unexpected cost to handle large numbers of invocations.

14.11 Summary

Serverless computing follows the traditional client-server paradigm in which one or more clients initiate contact with a server. Unlike a traditional concurrent server that is limited to one computer, serverless computing uses cloud technologies that allow it to scale arbitrarily. Also unlike traditional server designs, serverless computing separates server management from the core function of a server, allowing cloud providers to offer services that deploy and operate servers for their customers. To achieve maximum flexibility, providers offer arbitrary scaling with no minimum charge, known as *zero to infinity* scaling.

The chief motivation for serverless computing lies in its economic benefits. Because a cloud provider handles the details of managing server deployment and scaling, a customer does not need to maintain staff with expertise. Because a cloud provider only charges for the computation actually used, a customer does not pay for idle VMs or servers.

In terms of implementation, serverless computing adopts and extends the architecture used for Kubernetes controller-based systems. Serverless systems use an event-based paradigm in which each change in the cloud system and each contact from a client becomes an event that is added to a queue. A dispatcher repeatedly extracts events and assigns them to worker nodes to handle.

Despite all the advantages, serverless computing has potential disadvantages. Unlike a conventional server that starts before clients initiate contact, serverless computing creates servers on demand, leading to a small delay. Unexpectedly high costs can arise from cascades of events and from microservices that divide computation onto small functions.

Chapter Contents

15

DevOps

15.1 Introduction

Previous chapters describe programming paradigms and technologies used in building and running software in the cloud. This chapter takes a different approach by describing a methodology that software teams use to build and deploy software. The chapter gives the motivation, describes the methodology, and explains why it is especially appropriate for cloud software.

15.2 Software Creation And Deployment

To understand methodologies for engineering software, one must understand a few basic ideas, starting with how software is created and deployed. Software development starts with a business need. An organization identifies how their business can benefit from new software. If no commercial software satisfies the need, the organization uses its staff to create new software. In theory, after an initial engineering effort, the software remains available indefinitely, leading to a two-step deployment:

> Step 1. An initial engineering effort creates, tests, and deploys a new piece of software, making it available for use
>
> Step 2. Authorized users invoke the software as needed, possibly over a span of many years

15.3 The Realistic Software Development Cycle

In practice, the simplistic two-step deployment described above only suffices for the most trivial computer programs. As Chapter 12 describes, large, complex pieces of code contain errors despite the best efforts of software engineers. Whenever an error is discovered, software engineers must change the code to correct the error and then deploy the new version.

In addition to repair of errors, other factors may require software to be changed. For example, most software systems depend on an underlying operating system and/or libraries. If the vendor changes the operating system or the libraries, the software may need to be changed (or at least recompiled) to work with the new OS. Even if the OS does not change, users may suggest new features, extensions, or other changes to improve the program. As a result, a realistic software development process involves continual update. Industry uses the term *software development cycle* to capture the idea. Figure 15.1 illustrates the concept of a software development cycle.

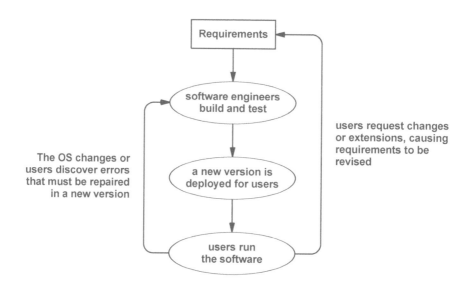

Figure 15.1 The concept of a software development cycle in which the discovery of errors or ideas for extensions and other changes require the software to be revised.

15.4 Large Software Projects And Teams

Because building and deploying software is complex, building and operating a nontrivial software system requires a team of people to work together. In fact, because the overall process involves a wide range of expertise, multiple teams each focus on one aspect. Most organizations divide the effort into three basic teams:

- Development team
- Quality Assurance team
- Operations team

Development team. The Development team consists of software engineers who start with a list of requirements, create a design, specify all the pieces, choose a programming language (or languages) and programming tools, and write the code.

Quality Assurance team. Before new software can be released to users, a *Quality Assurance (QA)* team performs extensive testing to ensure the software fulfills the requirements and works correctly.

Operations team. After testing the new software and certifying that it meets the requirements, the Quality Assurance team passes the software to an Operations team. The Operations team handles deployment, including configuring the set of users authorized to use the software, network addresses, and related details. The operations team also has responsibility for monitoring the execution, restarting after failure, planning backups, and creating multiple instances to handle scale.

15.5 Disadvantages Of Using Multiple Teams

Dividing the effort among multiple teams has the advantage of allowing each team to focus its effort and expertise on one particular aspect of software production. However, the separate team approach has disadvantages, including:

- Conflicting management objectives
- Competition and finger pointing
- Long development cycle

Conflicting management objectives. Each team has a manager who sets objectives and rewards for the team, and the rewards may differ. For example, the manager of the Development team might emphasize rapid development by offering a bonus for delivering software quickly, and the manager of the Quality Assurance team might emphasize taking time to be thorough by offering a bonus for finding errors.

Competition and finger pointing. The division into tasks inevitably leads to competition among teams and a tendency to blame other teams when problems occur. For example, if they find the software difficult to deploy and manage, members of the Operations team may blame the Development team for a poor design. Similarly, if the Operations team requests modifications or improvements, the Development team may blame them for delaying release of the software.

Long development cycle. Using a separate team for each part of the software project means each team waits until its task has been completed before passing the project on to the next team. With no overlap among teams, the total time from start to full operation can be months. We say that the long cycle results in a *big bang* (i.e., a major release) rather than a set of small, incremental improvements.

To characterize the interactions among teams in a traditional software development process, industry says that a team completes its task and then *throws the project over the wall* to the next team. Figure 15.2 illustrates the idea.

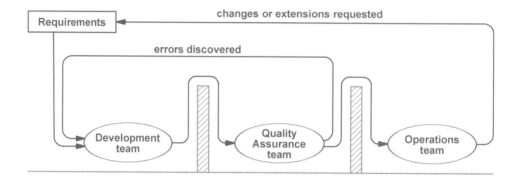

Figure 15.2 The flow of a software project across the conceptual walls separating teams in a traditional software development environment. Each team completes its task before throwing the project over the wall to the next team.

Although it may seem trivial, the notion of conceptual walls separating teams that each handle one step of the software process explains a key idea:

> *The traditional approach to software development in which a team completes its task before passing the project on to the next team can result in a long delay between a change being initiated and the deployment of a new version of software that incorporates the change.*

15.6 The DevOps Approach

For decades, software engineers have looked for ways to reduce the time required to create, test, and deploy new versions of software. A methodology known as *Agile* (sometimes written *agile*) focuses on self-organizing teams that work closely with users during planning and development, flexible and rapid responses to changes, continual and incremental software creation and improvement, and early delivery of software, even before all features are available.

The *DevOps* methodology adopts and extends the Agile approach. Instead of merely concentrating on building software applications that will be sold to customers who run them, DevOps focuses on software that will be deployed and managed by a local Operations team. Consequently, as the name implies, DevOps squeezes the three traditional teams (Development, Quality Assurance, and Operations) into a coordinated effort that works on all aspects of planning, building, testing, deploying, and operating software systems.

Including the Operations team in the software creation process helps ensure that the resulting software can be managed. For example, suppose the operations team intends to deploy the software using serverless computing. As we have seen, serverless computing requires server software to be stateless. Thus, the Operations team can inform developers about the intent, resulting in a stateless design. Similarly, an Operations team can advise the developers about the management of microservices and help developers plan a granularity that can be managed. Finally, the operations team can collect statistics about the performance of the software, and help developers improve successive versions.

15.7 Continuous Integration (CI): A Short Change Cycle

In a traditional software process, a long update time encourages developers to collect as many changes as possible into each release. Doing so further extends the time between releases. Industry uses the term *major release* to characterize a new version of the software that incorporates many changes.

Like the Agile methodology, DevOps espouses a philosophy of *Continuous Integration* (*CI*) in which the development team makes small changes in the software continuously rather than waiting to collect changes into a major release. Two ideas help make continual integration efficient and effective:

- The use of advanced development tools
- Continual testing of small pieces

The use of automated development tools. In a traditional software process, programmers write new code, place the code in a source repository, and build a new version of the program. Advanced development tools reduce the time required for development by automatically building a new version of the program whenever a programmer changes a piece of the code. Of course, the tools manage versions carefully to keep work-in-progress separate from the current production version.

Continual testing of small pieces. To achieve high quality, changes must be tested before new versions of software can be placed into production. To make rapid testing practical, software must be divided into small pieces because exhaustive testing of a large, monolithic application can take days or weeks. That is, disaggregating software into smaller pieces (i.e., following the microservices approach) helps reduce the time required to test each change. The point is,

To accelerate the software development cycle, DevOps divides software into small pieces that can be changed and tested quickly.

15.8 Continuous Delivery (CD): Deploying Versions Rapidly

In a traditional environment, the Operations team chooses a specific time and date to deploy each major software release. To minimize disruption, the team schedules a release to occur when use is minimal, such as at night or on a weekend. The change from one version to another can be abrupt: the operations team must terminate the old version before making the new version available. Users are typically warned in advance of a new release, allowing them to plan for the change.

The idea of scheduling a time for each major release does not work well in a cloud environment. To understand why, think of releasing a new version of microservice *M*. Many other services may depend on *M* being available. If the operations team stops all instances of *M*, the action may have a ripple effect that also stops critical services. Thus, stopping software before a new release may be inconvenient or impossible.

Fortunately, most cloud software uses containers. Because it exits once it handles a request, a container disappears quickly. Therefore, an operations team can allow old instances of the software to continue until they exit, and phase in the new version by using it for new instances. Because an organization's business functions depend on software being correct, new versions of software cannot be deployed without careful testing. By breaking software into smaller pieces, the microservices approach makes testing easier and faster. Because it has fewer parameters, a given module can be changed, tested, and then deployed rapidly; the approach works especially well for SaaS services (e.g., Office 365) because the SaaS vendor controls the release of new versions.

15.9 Cautious Deployment: Sandbox, Canary, And Blue/Green

Even though a module has been tested, most organizations proceed cautiously. That is, the organization does not merely deploy the new version into production. Instead, the organization may choose deployment strategies, such as:

- Sandbox trial
- Canary deployment
- Blue/Green deployment

Sandbox trial. One deployment approach creates a completely separate copy of the production environment for a trial deployment. As the term *sandbox* implies, the environment is isolated from the production environment. Thus, the new module can be tested to see that it interacts correctly with other modules without any risk to the production system. Duplication means the sandbox approach incurs high cost. Isolation from the production environment may mean a sandbox trial may not catch all errors.

Canary deployment. The *canary* approach overcomes some of the disadvantages of the sandbox approach. Canary does not use a separate environment for testing. Instead, the production environment is divided into two parts. Most users continue using the old software, and a few "canary" users are switched to the new version. Like a canary in a coal mine, the canary users serve as an early warning in case problems arise in the new version. Once the canary deployment succeeds, all users are switched to the new version.

Blue/Green deployment. Like the canary approach, the blue/green approach divides the production environment into two parts, known as *blue* and *green*. The blue part is reserved for testing the new version, while the green part continues to run the old version. Once testing completes, the two parts are reversed, with the blue version becoming the production version, and the green version used to test the next version.

Although the brief descriptions above make deployment strategies seem straightforward, many details complicate the implementation. The facilities must be managed carefully to ensure that a new software module will not produce incorrect results or cause harm. If a problem occurs, it must be possible to roll back to a previous, stable version quickly. More important, the system must be able to react quickly (e.g., in milliseconds). To summarize:

> *Because it involves many details and requires rapid reactions when problems arise, continuous deployment requires a separate tool to manage versions, testing, and trial deployments, and to roll back to a previous version if a problem arises.*

15.10 Difficult Aspects Of The DevOps Approach

Although it offers the advantages of rapid and continuous integration and deployment, DevOps does not offer a magical and painless way to improve software development. Some of the downsides include:

- Change of culture and rewards
- Higher risk resulting from rapid deployment
- System-wide failures as opposed to module failures

Change of culture and rewards. As we have seen, moving to DevOps requires a complete change in organizational structure and work culture. Instead of separate teams and separate management for development, testing, and operations, DevOps requires the groups to work together. Instead of rewarding competition, DevOps rewards cooperation. It can be difficult for managers and engineers to embrace a major change in their value system and adjust to the new approach.

Higher risk resulting from rapid deployment. Despite the safeguards and careful testing, any approach that emphasizes rapid deployment of new software versions increases the risk of introducing problems. One approach to managing risk consists of making changes in two steps. The first step introduces additions and support functions the new version will need, and the second step makes the change to use the support functions.

System-wide failures as opposed to module failures. To enable rapid deployment of new versions, DevOps (and Agile) focus on testing small modules. An important software principle declares that two failure modes can occur: an individual module or a system of many modules. An individual module fails if it does not correctly compute the intended function. Surprisingly, the interactions among modules can cause a system to fail even if all modules perform correctly. For example, a *deadlock* can occur if modules have circular dependencies. Dividing software into smaller modules helps streamline development by making it easier and faster to test a given module. However, increasing the number of modules increases the interactions among them, making it more difficult to test interactions.

15.11 Summary

The traditional software engineering paradigm collects many changes into each version of software, and then schedules a major release that puts the new version into production. Traditionally, software production uses three separate teams for development, testing, and operations. The traditional approach does not work well in a cloud environment because it is difficult to schedule a time to stop all instances and switch to a new version.

To speed the software cycle, the DevOps approach divides software into small modules that can be tested quickly and combines testing with development. To ensure that the resulting software can be deployed and managed, DevOps integrates the operations team with the development and testing teams. DevOps strives for continuous integration, allowing changes and improvements to occur quickly and continually.

Rather than collecting changes into major releases, DevOps strives for continuous deployment of new versions that each contain a minor change. Several techniques enable continuous deployment. The sandbox approach uses a separate copy of the environment for testing. The canary approach divides the production environment into two pieces, and allows a few users to try the new version while other users continue with the old version. The blue/green approach divides the production into two parts, using one to test the new version and another to run the old version. Once the new version has been tested thoroughly, all users move to the new version until a next version has been created.

Continuous deployment involves many details, including directing each user to an instance of the correct version and automatically rolling back to a previous version if a new version causes problems. Consequently, a separate tool may be needed to manage the deployment of new versions.

Despite its advantages, DevOps has downsides. It requires a complete change of the work culture and reward system, has higher risk of problems resulting from the rapid deployment of new versions, and the emphasis on smaller modules makes it more difficult to assess system-wide failures that occur from interactions among modules.

Part V

Other Aspects Of Cloud

Edge Computing And IIoT
Security And Privacy

Chapter Contents

16

Edge Computing And IIoT

16.1 Introduction

Previous chapters explain conventional cloud data centers. The chapters describe various aspects of cloud computing facilities, including infrastructure, virtualization technologies, orchestration systems, and programming paradigms. In a conventional cloud environment, computing, storage, and most communication occurs within each data center.

This chapter considers an alternative design known as *edge computing*. Instead of concentrating cloud facilities in a single geographic location, an edge architecture places some computational facilities near the source of data. The chapter gives the motivation for edge computing, and explains why the Industrial Internet of Things (IIoT) works well with edge processing.

16.2 The Latency Disadvantage Of Cloud

Recall from Chapter 1 that the move to a private cloud reverses a long-term trend in computing and moves back toward a centralized model. Public data centers represent a larger move toward centralization because a given cloud data center houses computing facilities used by many organizations. Although the cloud approach has many advantages, it does have the disadvantage of introducing higher network latency because a data center is remote from the customers it serves. As we have seen, cloud providers attempt to minimize network latency in two ways:

- The use of multiple, geographically diverse sites
- Low-latency network connections

The use of multiple, geographically diverse sites. To reduce latency, a public cloud provider does not collect all its facilities into a single data center. Instead, a provider creates multiple data centers, and places them at geographic locations (sometimes called *zones*) near sets of customers. For example, a provider might spread multiple data centers across North America, Europe, and so on.

Low-latency network connections. The second technique providers use to minimize latency involves low-latency network connections. A large enterprise customer may choose to lease a direct connection from the customer site to a cloud data center. To minimize latency for smaller customers, a major cloud provider connects directly to Tier-1 Internet backbone networks.

Despite the above optimizations, the latency between a user and a cloud data center can exceed 100 milliseconds, and may depend on network traffic. The point is:

> *Although public cloud providers employ optimizations to minimize latency, the cloud model introduces a delay between a customer and an application running in a cloud data center.*

16.3 Situations Where Latency Matters

Is latency important? In many situations, no. When a corporation performs routine business (e.g., recording sales transactions or submitting monthly payroll information), a slight delay is unnoticeable. In some situations, however, low delay can be crucial, either financially or otherwise. In the financial industry, for example, a small delay in making stock trades can result in a huge loss. In the health care industry, a small delay in receiving data from a patient monitor can delay activation of an implanted medical treatment device.

The point is:

> *Although the cloud approach works well for traditional computation, the network latency incurred when sending data between a customer and a remote data center can make cloud computing inappropriate for some applications.*

16.4 Industries That Need Low Latency

Consumers seldom worry about low latency. For example, when a user interacts with the controls in their smart home (e.g., to change the temperature), small delays go unnoticed. However, industries that employ real-time control systems —sensors and actuators that monitor and control processing —rely on low delay. Systems that provide fast responses to human users (e.g., responses to keystrokes a user enters) also need low latency. Figure 16.1 lists some of the industries for which low latency can be important.

Agriculture	Mass transit
Automotive	Oil and gas
Environmental monitoring	Retail
Health care	Transportation
Manufacturing	Utilities

Figure 16.1 Example industries in which applications require low latency.

Of course, not all sensors require low latency. Even some medical devices do not require instant responses. For example, wearable medical devices exist that collect biometric data for long-term trend analysis, either by an AI program or a human medical professional. Such devices typically accumulate measurements over multiple hours or multiple days before uploading values to the cloud.

16.5 Moving Computing To The Edge

How can cloud computing be adapted to meet the requirements for low latency? The answer lies in an architecture known as *edge computing*. The idea is straightforward: place some of the computing facilities near each source of information, and perform initial processing locally. Simultaneously run applications in a cloud data center, and use the cloud applications to handle computational-intensive tasks. The term *edge* arises because cloud data centers typically connect to a centralized point of the Internet, whereas networking professionals say that users' devices connect to the "edge" of the Internet. Hence, computation performed near such devices occurs near the edge.

To understand how local computing can help, consider a simplistic case: a sensing device. Suppose an app running on a user's cell phone uses Bluetooth technology to collect data from a sensor and then sends the data to an application running in the cloud for analysis. Figure 16.2 illustrates the idea.

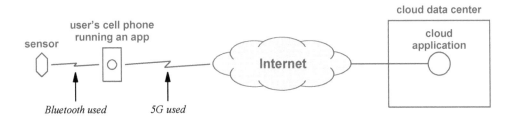

Figure 16.2 An app on a user's cell phone that gathers information from a sensor and sends the information to an application running in a cloud data center for analysis.

The application running in the cloud has access to powerful computing facilities, and can perform computationally intensive processing, such as running AI software to analyze the sensor data. Meanwhile, the app on the user's cell phone handles local tasks. The app uses Bluetooth to communicate with the sensor and gather data. It can also perform basic data processing before sending the data on to the cloud. For example, electrical interference (e.g., from lightning) can temporarily disrupt communication with the sensor. In such cases, the data being transferred from the sensor may be lost or corrupted. The app running in the cell phone can detect such problems, discard corrupted data values, and restart the transfer once the interference passes. In addition, the app can inform the user if the sensor becomes disconnected or the battery in the sensor needs to be recharged.

To expand local computing capabilities, the edge computing approach places a miniature data center near locations that require low latency responses. Software running in the edge data center performs computation that requires low latency, and software running in a cloud data center performs computations that do not require rapid responses. The point is:

> *The edge computing approach places small, auxiliary data centers near locations that require low latency responses. Software in the edge data center handles low-latency computations locally and runs other computations in a cloud data center.*

16.6 Extending Edge Computing To A Fog Hierarchy

Where should edge data centers be placed? The locations and sizes depend on the applications being supported and the latency requirements. To achieve the lowest possible latency, an edge facility must be as close to each user as possible (e.g., in each cell tower). For applications with less stringent requirements, an edge computing facility might serve a neighborhood of multiple cell towers or a geographic region with many neighborhoods. To serve all applications, edge facilities can be organized into a hierarchy. Figure 16.3 illustrates one possible arrangement.

Level	Computing Equipment	Connects To Multiple
1	Public cloud data center	Regional data centers
2	Regional data center	Town data centers
3	Town or neighborhood data center	Cell towers
4	Computers in a cell tower	Users' phones
5	User's phone	Sensoring devices

Figure 16.3 A possible edge computing hierarchy with computing facilities arranged into five levels.

To distinguish between edge facilities located adjacent to end users and edge facilities that serve larger geographic regions, industry sometimes uses the term *fog data center* to refer to an intermediate data center that serves a larger geographic area. The name is intended to be humorous by referring to the idea that fog arises when a cloud comes down to earth. By analogy, a fog data center uses cloud technology, but places a smaller data center closer to endpoints.

To summarize:

> *Industry reserves the term* edge data center *for a small data center directly adjacent to endpoints, and uses the term* fog data center *to refer to an intermediate data center in an edge hierarchy.*

16.7 Caching At Multiple Levels Of A Hierarchy

Although the description above focuses on computation, a multi-level hierarchy of edge and fog data centers permits another optimization: data caching. When an endpoint requests information, an application running in the nearest edge data center can obtain the requested information, store a copy locally, and return a copy to the requester. If another endpoint subsequently requests the same information, a copy will be returned from the edge data center without any need to contact the original source.

Caching can be used at all levels of the hierarchy, and works for data flowing in either direction. For example, suppose a hierarchy contains a cloud data center at the top, two fog data centers that each serve a region, and edge data centers in each region, as Figure 16.4 illustrates.

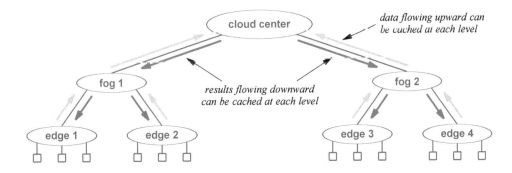

Figure 16.4 An example hierarchy of edge and fog data centers that each cache copies of data.

As endpoints generate data, the data flows upward, and each data center keeps a cached copy. Thus, if an endpoint connected to edge data center *edge 1* generates data, *edge 1* caches a copy and forwards the data to *fog 1*, which keeps a cached copy and forwards a copy to the cloud, which also keeps a copy in its cache. If another endpoint connected to edge 1 requests the data, *edge 1* will return the values from its cache. Similarly, if an endpoint attached to *edge 2* requests a copy, *edge 2* will obtain the data from *fog 1*, cache a copy and return a copy to the endpoint that made the request.

As the figure indicates, caching also applies to data items flowing down the hierarchy, and can include items that span multiple applications. As an example, consider a supply chain management system for a retailer with multiple stores. As items are sold, information is propagated upward through the hierarchy, allowing regional managers to order replacement items from a supplier. The system also maintains information used to ship items between stores to obtain items from a nearby store when the supply temporarily runs low. Such information originates from a central company database and flows downward through the hierarchy, as needed. The caching system works equally well with both types of information.

To summarize:

> *In addition to performing computations, a hierarchy of edge and fog data centers can cache arbitrary information at each level. Caching lowers latency when the system receives multiple requests for a given data item.*

16.8 An Automotive Example

A single application may benefit by using multiple levels of the hierarchy. To see how, consider the automotive industry, which is working to create a system to support *connected vehicles*†. Once the system becomes operational, each vehicle, whether self-driven or driven by a human, will communicate with nearby vehicles as well as with communication facilities permanently placed near roadways. The system will handle safety as well as navigation. For safety, mechanisms will ensure that a vehicle will not collide with another vehicle nor cause problems for others by maneuvering dangerously or slowing needlessly.

Three aspects of the connected vehicle system lend themselves to the edge computing approach.

- Low latency / real-time requirements
- Geographic locality and awareness
- The wide scope needed for route planning and navigation

Low latency / real-time requirements. Because vehicles travel at high speed, collision avoidance systems operate locally. Such systems maintain information about the

---

†See https://aecc.org/ for information about the *Automotive Edge Computing Consortium* (*AECC*).

position and direction of surrounding vehicles, allowing them to calculate actions quickly when a problem occurs.

Geographic locality and awareness. To calculate a safe speed and choose an action that will avoid a collision, a collision avoidance system must also have information about local road conditions. For example, is the roadway rough or has it become covered with ice? Similarly, a vehicle must be informed when an accident closes the road or when traffic ahead stops unexpectedly. Because they must act quickly, vehicles closest to the problem must learn of the problem first. That is, notification mechanisms must be aware of geographic positions.

The wide scope needed for route planning and navigation. Route planning systems use map data to perform a global optimization: the system chooses a path that minimizes the travel time over the entire trip. However, navigation systems must also adapt to local conditions, routing around temporary problems. Thus, route planning and navigation require information about local conditions.

The items described above suggest how automotive systems might use a hierarchy of edge computing facilities. A vehicle can track adjacent vehicles directly. To report current conditions and to learn about traffic and the roadway ahead, a vehicle might communicate with edge facilities along the road (e.g., in each cell tower). The edge facilities can learn about changes and report them rapidly (in seconds).

At the next level of the hierarchy, we can imagine a set of data centers that each serve a small area. The data center runs software that collects and correlates information from multiple edge facilities. For example, by receiving reports of rain on various roads, software can calculate the path of a rain storm as it moves through the area and warn vehicles before they encounter the storm. Information in such data centers tends to change slowly (e.g., in minutes instead of seconds).

At the third level of the hierarchy, we can imagine a set of regional data centers that each collect and process information from areas within their region. For example, systems in a regional data center might use AI to analyze traffic patterns and discover that a particular route becomes congested between 4:00 PM and 7:00 PM on workdays.

The point is:

The envisioned system for connected vehicles illustrates how a hierarchy of small edge and fog data centers can provide low-latency responses and manage information over a range of geographic areas.

16.9 Edge Computing And IIoT

The term *Industrial Internet of Things* (*IIoT*) refers to an enhanced, larger-scale version of the Internet of Things. The primary difference between consumer IoT systems and Industrial IoT systems lies in the importance: a company depends on an IIoT system as part of a critical business function.

As an example of IIoT, consider automated manufacturing. The assembly line in an automated factory consists of robots at each station plus conveyors that move items down the line. Raw materials enter at one end, and finished products emerge at the other end. Because such assembly lines form a critical aspect of the company's business, the company loses money if the line stops. Consequently, sensors at each stage of the assembly line monitor items entering the assembly line, robots along the line, the progress of the entire line, and the quality of the items being manufactured. A control system gathers data from the sensors, analyzes the data to detect and predict problems, and responds either by resetting equipment or notifying a human when it cannot handle a problem.

An automated assembly line illustrates the characteristics often found in IIoT applications that distinguish them from most consumer IoT applications:

- Specific latency requirements
- Geospatial knowledge
- Large volumes of data with various QoS requirements
- A need for data filtering
- High availability requirements
- Security requirements

Specific latency requirements. Instead of a general desire for high performance, IIoT applications have specific requirements (e.g., if a specific assembly line robot malfunctions, it must be shut down within 150 milliseconds after a problem is detected).

Geospatial knowledge. An IIoT system must be aware of locations and spatial relationships (e.g., the location of a failure and the set of surrounding systems that will be affected).

Large volumes of data with various QoS requirements. IIoT applications often employ many sensors and video cameras that each generate data continuously, resulting in large volumes of data; technologies such as 5G wireless enable especially high traffic volumes. Each type of data may have specific *Quality of Service* (*QoS*) requirements, such as data rates and bounds on latency.

A need for data filtering. It does not make sense to send all the raw data gathered from sensors to the cloud for processing because data transport and computational cycles incur expense. More important, because a local edge data center can handle items that require immediate action, applications running in the cloud only need data that allows them to analyze long-term trends (e.g., whether a given factory has more failures than other factories). Even local processing benefits from filtering (e.g., because an adaptive cruise control application only handles sensor readings for nearby objects, filtering avoids having the application waste cycles on other data).

High availability requirements. Because a company depends on IIoT systems to sustain their business, the systems must be reliable. Thus, IIoT may need to employ redundancy (e.g., have multiple sensors monitor a given piece of equipment in case one fails and reports incorrect values).

Security requirements. IIoT systems must be secure from attack, and it must be possible to keep the data they gather confidential. For example, a robot should not accept a command that is not authenticated, and a biometric sensor should not send medical data over a network until the data has been encrypted.

We can summarize:

> *An IIoT system may transfer multiple types of data, each with its own requirements for performance, availability, and security.*

16.10 Communication For IIoT

A typical IIoT application involves many sensors generating data and multiple applications running in a hierarchy of edge and fog centers processing the data and (possibly) issuing commands that control the underlying devices. As the previous section points out, the system must meet specific requirements for Quality of Service. The question arises, "How should communication be arranged among all the pieces to achieve the desired goal?" One answer comes from the *Object Management Group (OMG)*†. An OMG standard known as the *Distributed Data Service (DDS)* defines a mechanism that allows data from sensors to flow upward through a hierarchy of edge and fog centers to applications using the data. The DDS approach has the following characteristics:

- Completely decentralized
- Suitable for industrial use
- Publish-Subscribe interactions
- Flexible data handling capabilities
- Support for an edge hierarchy

Completely decentralized. Unlike communication systems that rely on a process to distribute data to subscribers, DDS avoids a single point of failure while minimizing latency by using direct communication. DDS can also use multicast transmission over a network to reach multiple applications efficiently.

Suitable for industrial use. DDS offers the high reliability needed for IIoT applications. It can be configured to meet performance requirements and to prioritize specific types of data. In addition, DDS offers the ability to authenticate control messages and encrypt data traffic.

Publish-Subscribe interactions. DDS offers a *publish-subscribe* communication mechanism that allows each application to choose the data the application will receive. An application running in the edge may choose to receive all data from a given sensor, and an application running in a fog center may choose to ignore the raw data and only receive periodic summaries.

---

†OMG is also the parent organization for the *Industrial Internet Consortium (IIC)*, a group of companies working to accelerate the creation and adoption of IIoT.

Flexible data handling capabilities. DDS provides flexibility that allows users to specify filtering and QoS on a per-interaction basis. For example, suppose a sensor system publishes data. One subscriber may choose to receive all data while another chooses to receive only values beyond a specified threshold. Each subscriber can also negotiate QoS requirements independently.

Support for an edge hierarchy. DDS can be configured to form a distributed system where some of the participating applications run in an edge data center and others run in fog centers at higher levels of a hierarchy. The hierarchy can extend to include public clouds as the highest level.

DDS uses the term *Databus* to describe the mechanism used to support publish-subscribe communication. Multiple applications connect to a Databus, and each application can choose to publish data and subscribe to data published by other applications. As the name implies, a Databus provides interconnections among publishers and subscribers analogous to the way a hardware bus in a computer provides interconnections among I/O devices, memories, and processors.

Unlike the conventional bus in a computer, a Databus is not a hardware mechanism. Instead, one can think of a Databus as a communication abstraction implemented by software. Software modules arrange to send messages between publishers and subscribers to meet requirements and optimize communication. In practice, Databus software sends messages across underlying computer network(s) and other computer communication channels.

Databus technology includes a mechanism known as a *gateway* that allows a single Databus to span multiple levels of the hierarchy. The system allows a given application to subscribe to data published at any level of the hierarchy, as if all applications connect to the same Databus. Rather than blindly sending a copy of every message, gateways perform a filtering function, and only forward a message to another level if one or more applications have subscribed to receive the message. As a result, the system provides the illusion of a single, large communication system, but eliminates unnecessary message propagation.

Figure 16.5 illustrates a Databus that uses gateways to span multiple levels of a hierarchy. In the figure, applications connect at each level. In addition, the figure shows databases that can be used to cache information.

16.11 Decentralization Once Again

Recall from Chapter 1 that computing started with expensive centralized mainframes, and that the advent of small, low-cost computers encouraged a move to a completely decentralized model with many computers connected by computer networks and the Internet. Also recall that cloud computing moved us back to a centralized model and increased centralization made it easier and less expensive to manage computing facilities. The idea is that at each stage, a new architecture overcame weaknesses in the previous architecture.

Edge computing adds a final ironic twist to the evolution of computing. Edge proponents point out that cloud computing has the weakness of high latency, and assert that a distributed architecture can overcome the weakness. Of course, moving back toward a more distributed architecture makes the system susceptible to some of the disadvantages that plagued earlier distributed architectures. In particular, like earlier distributed systems, a hierarchy of edge and fog data centers has the disadvantage of being more difficult to manage than a centralized cloud data center.

The point is:

> *In an ironic twist, edge computing moves away from centralized clouds, which arose to overcome the weaknesses of distributed computing, back toward a distributed model.*

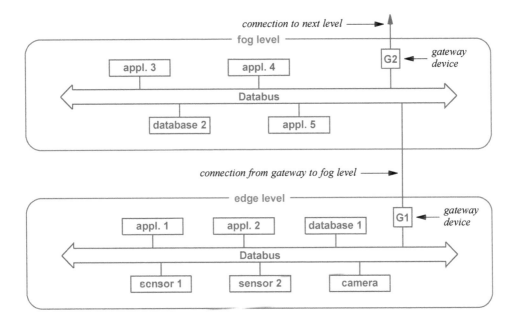

Figure 16.5 The concept of a single DDS Databus that spans multiple levels of a hierarchy. An application at any level can subscribe to receive data from an application at an arbitrary level.

16.12 Summary

To overcome the disadvantage of high latency, edge computing places small data centers close to users. Applications in an edge data center handle processing that requires low latency, and applications in a cloud data center handle other processing, especially long-running, computationally-intensive tasks. The concept of edge computing can be extended to a hierarchy of fog data centers, with successive levels of the fog hierarchy serving successively large geographic regions.

Some applications only need one level of an edge hierarchy. The automotive industry effort to design systems for connected vehicles illustrates that some applications can use multiple levels effectively. For safety and navigation, vehicles need to learn about changes in local road and traffic considerations quickly. For longer-term planning and navigation, a vehicle needs to obtain information about conditions on roads farther away.

Although it can provide convenience for consumers, edge computing is essential for Industrial Internet of Things (IIoT) applications that have high data volumes, a need for data filtering, and strict requirements for latency, reliability, and security. As an example, sensors in an automated factory generate a high volume of data and require rapid responses when a sensor reports a problem.

The Object Management Group offers a standard for communication among IIoT systems known as the Distributed Data Service (DDS). DDS supports publish-subscribe interactions among applications, has suitable reliability and security for a hierarchy of edge and fog data centers, and allows authentication and encryption. Using DDS, an application can subscribe to receive specific data items and specify security and QoS constraints.

DDS uses the term *Databus* to describe a conceptual communication mechanism that connects applications and implements publish-subscribe interaction. A Databus consists of software modules that implement publish-subscribe interaction by sending messages across underlying networks.

Conceptually, a single Databus can span multiple levels of an edge and fog hierarchy. Physically, the DDS standard defines a gateway device used to connect the Databus in one level to the Databus in another level. Once gateways have been created and configured, applications at multiple levels can subscribe to any published data as if they all connected to a single, large communication mechanism.

Edge computing reverses the move toward centralizing computing in cloud data centers by returning to a more distributed model of computing. Edge provides lower latency, but has the disadvantage of making management more difficult.

Chapter Contents

17

Cloud Security And Privacy

17.1 Introduction

Previous chapters describe cloud infrastructure, technologies, and the design of cloud-native software. The chapters also describe many of the advantages of cloud computing. This chapter considers a disadvantage: security vulnerabilities that arise in a cloud environment.

Of course, cloud computing has the same security vulnerabilities as conventional IT infrastructure. Our discussion focuses on the aspects of cloud that make conventional approaches to security inadequate or ineffective, and describes some of the techniques cloud users can employ to keep their data safe and confidential.

17.2 Cloud-Specific Security Problems

How does the use of cloud make it difficult for an organization's IT staff to manage security and privacy? Several factors contribute to increase the complexity of managing cloud computation, communication, and data storage.

- Lack of control and visibility
- An infrastructure shared with outsiders
- Many services with interdependencies among them
- Dynamic execution environment with bursts
- Remote access for all users
- Extensive use of software from the cloud provider and third parties

Lack of control and visibility. In a traditional IT environment, the staff controls the hardware and software facilities, and can investigate the root cause when problems arise. In a cloud environment, however, a tenant cannot configure or examine the underlying systems, and must trust that the provider's staff has configured security protections correctly. When a problem arises, the tenant must rely on the provider's staff to diagnose the cause and affect repairs. Furthermore, if a tenant's services do not perform as expected, the tenant cannot easily tell whether the problem lies in the underlying cloud infrastructure, the way the tenant has designed and deployed systems, or interference caused by an attacker.

An infrastructure shared with outsiders. Unlike a traditional IT infrastructure that serves one organization, multiple tenants share cloud infrastructure. In theory, virtualization technologies used for computation, communication, and storage provide isolation between a given tenant and other tenants. In practice, however, running multiple tenants on the same infrastructure increases security risks, and security and privacy breaches have occurred.

Many services with interdependencies among them. As we have seen, cloud systems encourage a microservices design in which many small services run independently with communication among them. In theory, a microservice can be protected from unauthorized access or attack as effectively as a traditional application. However, having many microservices and allowing frequent communication among them increases the attack surface, giving attackers more opportunities to find vulnerabilities.

Dynamic execution environment with bursts. Orchestration systems achieve elasticity by expanding services as needed, with new instances spread across multiple physical servers. Such systems often exhibit bursty behavior in which many new instances appear within a short span of time. Rapid creation of new instances makes it more difficult to distinguish normal execution from a *Denial of Service* (*DoS*) attack.

Remote access for all users. In a traditional IT setting, both the facilities being accessed and the employees accessing the facilities reside in a single location. Cloud computing reverses the situation because a tenant's employees, whether working in their organization's offices or at home, must use a remote access mechanism. In theory, security mechanisms protect remote access and ensure safety. In practice, the use of remote access increases the possible attack surface, especially in cases where responsibility for installing and managing security software falls to employees using their own devices.

Extensive use of software from the cloud provider and third parties. Perhaps the most significant security weakness in cloud systems has arisen from a major shift in software. A traditional IT department either purchased software from well-known (and trusted) suppliers or hired software engineers to create custom software from scratch. Building cloud-native software has added complexity, such as designing microservices that can be orchestrated. As a result, many cloud users now incorporate pieces of open source software into their systems. For example, container software can be downloaded from Docker repositories or GitHub. One publication describes such repositories as a "wild west" in terms of security.

The point is:

> *Although cloud computing provides many benefits, cloud introduces new security risks. Vulnerabilities include: dependence on the provider to configure protection, sharing infrastructure with outsiders, increased attack surfaces caused by microservices, a dynamic execution environment, remote access, and the extensive use of open source software obtained from public repositories.*

17.3 Security In A Traditional Infrastructure

To understand the motivation behind some of the new security techniques being used in cloud systems, one must understand traditional approaches to security and see why the traditional techniques do not work well in a cloud environment. We will examine four aspects of traditional security.

- Insiders vs. outsiders
- Perimeter security
- Demilitarized zones (DMZs)
- Standing privileges divided into a few levels

Insiders vs. outsiders. A traditional IT infrastructure divides people into major groups: *insiders* and *outsiders*. Insiders include employees and contractors who work for the organization. Temporary insider status may be granted to IT vendors when they need to install or repair hardware and software systems. The infrastructure classifies anyone else who tries to access the organization's IT facilities as an outsider. Although the terminology implies physical proximity, the classification arises from the identity of the individual, not whether they gain entry to the organization's buildings.

In a traditional IT environment, all aspects of the system must be designed to support the insider/outsider distinction. For example, a typical organization runs two Wi-Fi networks in its buildings: a main campus network that insiders use to conduct company business and a *guest network* that visitors and other outsiders can use to access the Internet and public company sites. To enforce security, the guest network remains completely isolated from the organization's main campus network.

Perimeter security. Traditional security systems follow the fortress approach by enclosing all the campus IT facilities with conceptual walls that secure the perimeter. That is, to prevent attackers from launching attacks from outside, the organization deploys security mechanisms at each connection to the outside world to control access. Example perimeter security mechanisms include *firewalls*, *Deep Packet Inspection* (*DPI*) systems, and scanners used to detect viruses and other malware in incoming email and imported files.

In essence, perimeter security extends the idea of insiders and outsiders by making it especially difficult for an attacker who does not have physical access to the organization's IT infrastructure. Interestingly, rather than create a cyber attack that can penetrate perimeter security, attackers often use a less technical approach by bribing employees and contactors to gain physical access inside the perimeter.

Demilitarized zones (DMZs). One particular perimeter security mechanism that prevents unwanted access consists of placing a *demilitarized zone* on each external connection. The idea is straightforward: instead of allowing arbitrary network traffic, restrict access to a specific set of servers. To enforce security further, place the servers on an isolated network. For example, an organization can use a DMZ to enable outsiders to access a public web server and the server that accepts incoming email without risking access to other networks and servers in the organization.

Standing privileges divided into a few levels. Traditional security uses a semipermanent assignment of privileges to individuals in a scheme known as *standing privileges*. That is, once an individual has been assigned a privilege level, the assignment remains unchanged from day to day.

A typical organization divides standing privileges into a few levels. In the simplest case, two levels suffice: one for users and the other for system administrators who are responsible for installing and operating IT systems. In a large organization, privileges may be compartmentalized among a set of individuals according to their duties (e.g., one group of IT staff members has privilege to manage database servers, while another group has privilege to manage web servers).

17.4 Why Traditional Methods Do Not Suffice For The Cloud

As we have seen, the algorithms that cloud providers use to allocate VMs and containers choose a location for each new instance that avoids hot spots. That is, instead of devoting a physical set of servers to a given tenant, a physical server may run VMs and containers from multiple tenants. More important, even if the provider uses virtual networking technology to give each tenant a separate virtual network, traffic from multiple tenants passes over the underlying physical network links, which means potential attacks may arise from within the facility. The lack of a physical perimeter has consequences: a tenant cannot divide potential users into insiders and outsiders, and cannot depend on perimeter control mechanisms to keep attackers out.

> In a cloud environment, a tenant cannot rely on traditional perimeter
> security to define the inside and outside of their network.

The use of microservices further complicates the cloud environment. To minimize risks, an organization must follow the *Principle of Least Privilege* (*PoLP*) and give an individual the least privilege the individual needs to perform their job. Because a given IT staff member will not need absolute privilege for all microservices, privileges are not merely divided into users and system administrators. Instead, a security system must allow each individual to be authorized for specific types of access to specific services. To summarize:

In a cloud environment, instead of granting broad access privileges, the security system must allow each individual to be authorized for specific types of access for specific services.

Unlike a traditional IT infrastructure, a cloud tenant cannot manage security on external internet connections directly. Instead, a tenant must work with the cloud provider. The tenant defines security policies, and the provider configures hardware and software systems that implement the policies. If a tenant changes its policies, the provider must adjust the configuration accordingly.

17.5 The Zero Trust Security Model

Consider *perimeterless security* (sometimes called *borderless security*). Without a perimeter, how can a tenant know which individuals should be allowed to access and manage services? The answer lies in a *zero trust security model*. The idea is straightforward: assign each user a set of privileges for each possible service. Instead of merely allowing a user to login once and then have access to all services, validate each request separately. That is, whenever a user attempts to access a service, use the identity of the user (and possibly the identity of the user's device) to decide whether to grant or deny the request.

Moving to a zero trust model can be difficult because it requires choosing a balance between security and convenience. As an example, consider a web page that an employee can use to find another employee's email address. Further suppose the lookup mechanism involves two services, one that allows a user to enter an employee's name and returns the employee's ID, and another that takes an employee ID as input and returns the employee's email address. A naive implementation of zero trust security checks the user's authorization at each step. First, the web page asks the user to enter login and password credentials. The web page then invokes the name lookup microservice, which also prompts the user to reenter the same credentials. Once the name lookup succeeds, the web page invokes the email lookup service, which prompts the user for credentials. A user can easily become frustrated with a system that asks the user to log in multiple times while performing a single task. The situation is much worse if each service maintains its own set of login IDs and passwords.

In practice, implementing a zero trust model requires a centralized mechanism that handles identity (i.e., passwords and authentication), as explained in the next section. Attempting to implement a zero trust model without such a system introduces security holes because it leaves each subsystem free to assign levels of privilege without coordination across all subsystems.

17.6 Identity Management

How can a set of microservices be designed to remain secure without using separate logins for each service and without requiring each invocation to check a user's credentials? The answer lies in a software system for *Identity Management (IdM)*. To avoid having separate authentication for each service, an identity management system uses a *Single Sign On (SSO)*, which means a user has the same login and password credentials for all services. The system stores information about each individual, including their login, password, and access rights. All services use the identity management system (which can be replicated to handle scale) to ascertain whether a given individual can access the service. Of course, accesses to the identity management system use encryption to ensure that the information is kept safe.

In addition to merely providing a way to check on access rights, an identity management system can be designed to return a digital *capability*. The capability can be passed to other services, allowing each service to determine whether a user is authorized to access the service without requiring the user to enter their credentials multiple times. As a result, an identity management system offers convenience for users, while assuring that a user's access to a service has been authenticated and the access has been authorized. The point is:

> *An Identity Management system, which stores information about users' identities and their access rights, uses a single login for all services, authenticates users, ensures only authorized users access each service, and allows a user to enter credentials once for each task.*

17.7 Privileged Access Management (PAM)

Controlling access for IT staff forms a special case of Identity Management. Because they need permissions to install, configure, and operate systems, IT staff members have the *administrative* or *superuser* level of privilege. An attacker who can forge the credentials of an IT staff member may be able to obtain confidential data or inflict serious damage on an organization's digital systems. Thus, an organization must take special care to guard such access.

A *Privileged Access Management (PAM)* system handles identity management for privileged accounts. One significant feature of PAM systems arises from their emphasis on limited privilege: instead of a master password that grants administrative privilege on any system, a staff member only has privilege on the systems the staff member administers. In addition to checking credentials and limiting access, PAM systems log all accesses, providing a detailed record of which individual accessed a system at a given time. The systems also record failed login attempts, providing a way to track attackers.

Some PAM systems provide further checks on the use of accounts. For example, when adding a staff member to the system, the manager configures a primary set of systems for which the staff member has full responsibility and a secondary set for which the staff member serves as backup when staff with primary responsibility are not available. If a staff member's ID is used to access a secondary system while a primary staff member also has access or a staff member's ID is used to access multiple secondary systems within a short time, the PAM system sends an alert to the manager reporting suspicious activity.

17.8 AI Technologies And Their Effect On Security

Interestingly, *Artificial Intelligence (AI)* technologies influence security in both positive and negative ways. On the one hand, attackers can use AI techniques to bypass safeguards and gain unauthorized access to data or systems. On the other hand, AI techniques can be used to strengthen safeguards.

As an example of attackers using AI, consider an attack where AI technology was used to fool an employee rather than to target the company's IT systems from the outside. The approach is known as a *deep fake*. An AI program that used *machine learning (ML)* was first fed recordings of the voice of a company executive so the program could learn the executive's voice and speech patterns. Once the learning phase was complete, the program imitated the voice while pronouncing text from a script. The attacker left voice mail for an employee, apparently from the company executive, asking the employee to transfer confidential data to an external site.

As a positive use of AI, recall that PAM systems, discussed above, can alert a manager if a staff member's ID is used to obtain privileged access to systems for which the staff member does not have primary responsibility. The idea has been extended by using AI software to detect various types of anomalous behavior under the general topic of *security analytics*. In essence, analytics software determines "normal" behavior, and then looks for exceptions. For example, an AI system may be able to spot unusual behavior by an employee, such as accesses made to data or services that the employee does not usually access or accesses that occur at unusual times for the given employee (e.g., access to a system early in the morning by an employee that only accesses the system after noon). Similarly, if a machine learning program receives notification of each network connection that arrives from the Internet, the program may be able to detect an instance of a *Distributed Denial of Service (DDoS)* attack.

One final aspect of security analytics arises from *context*. Suppose a given user always logs into the HR system and uses the system to update the employee database. If the user accesses the employee database directly, the identity management system will grant access because the user is authorized for such access. However, security analytics software takes another view: it flags the access as unusual because it does not occur in the usual context. We say that security analytics software can be *context-aware*.

The point is:

> *Although they enable new types of attacks, AI technologies can also be used to increase the effectiveness of safeguards.*

17.9 Protecting Remote Access

Many organizations permit employees to access the organization's traditional IT infrastructure from remote locations. When an organization moves to cloud computing, remote access becomes the default mode of interaction. A further complication arises with remote access because many organizations now allow employees to use their own devices and to have multiple devices (e.g., a phone and a laptop).

Many questions arise about remote access. How can an organization keep data and computations safe if employees can download items from the cloud onto their own devices? How can an organization ensure that communication sent over the Internet between an employee's device and the cloud data center remain confidential?

Remote access extends the notion of borderless security by extending the boundary outside the cloud data center. Despite the physical location of remote devices, an organization must still maintain a boundary between items that must remain secure and threats to those items. Three principles help define security practices:

- Keep all communication confidential
- Protect and isolate business data
- Enforce workflow security

Keep all communication confidential. When an employee uses the Internet, steps must be taken to prevent outsiders from eavesdropping. For example, consider an employee who uses a Wi-Fi network, whether in their home or in a public location, such as a coffee shop. An outsider can easily capture all the packets sent from or to the employee's device.

To keep communication confidential, all data must be encrypted. Fortunately, software exists that solves the problem. Known as a *Virtual Private Network* (*VPN*), the technology forms a connection to the organization's cloud data center and encrypts all packets sent over the connection, thereby preventing outsiders from interpreting the contents of packets. Furthermore, to prevent accidental leaks between normal communication with Internet sites and communication with the organization's servers, once a VPN has been invoked, it sends *all* packets to the cloud data center, even packets eventually destined for some other site on the Internet. Thus, the organization can impose restrictions on Internet communication.

Protect and isolate business data. Remote access introduces an additional danger: an employee may lose a device that contains confidential business data. Many instances exist where employees accidentally left their laptop or cell phone in a taxi cab, airplane, or other public location.

To protect it from accidental loss, all business data stored on a user's device must be encrypted. Fortunately, data encryption technologies exist that make it easy to enforce confidentiality and keep data safe. Some operating system vendors offer whole disk encryption schemes that can prevent outsiders from reading data without the correct credentials.

Enforce workflow security. As an employee performs a task, data may move from the cloud to the employee's device and back. Each set of data will have specific requirements. It may seem that the problem can be solved by choosing the most stringent set of requirements and enforcing the requirements on all data. However, doing so imposes unnecessary overhead because many restrictions only apply to a few cases.

To ensure that all data remains safe independent of its location, an organization must define a security policy for each *workload*, and keep the policies enforced at each step as the data moves. That is, the security policy for data is bound to the data itself. Industry uses the term *workflow security* to characterize the approach.

We can summarize:

> *Because cloud systems rely on remote access, an organization must take care to protect business communication and business data as the data moves to and from employees' devices.*

17.10 Privacy In A Cloud Environment

Security systems enforce protections to guarantee the confidentiality, integrity, and availability of data. In addition, organizations must comply with regulations that require special treatment for some forms of data (e.g., in the US, HIPAA regulations apply to the collection, storage, and transmission of data containing individuals' medical information). We use the term *privacy* to refer to keeping sensitive information about an individual safe from public dissemination.

It may seem that privacy can be achieved merely by keeping each individual's data confidential. However, privacy can be more subtle because it may be possible to deduce information about an individual from statistics, even if the actual data for the individual remains confidential. For example, consider a graph that shows the average income of a sample of people. Suppose the graph divides the group by age and shows the average income for those in their 20s, 30s, 40s, and so on. Further suppose the sample is so small that some categories only contain a single person. If an outsider can find the age of people in the sample, the outsider could deduce the identity, and therefore the income, of individuals who happen to be in a category of one person.

Privacy protection cannot focus only on the data at hand. Because the Internet contains so much information about individuals, it may be possible to identify individuals by combining data and statistics from many sources. Thus, the difficult question becomes, "Will publishing new information such as aggregated statistics make it possible for outsiders to deduce private information about individuals?" Colin Bennett wrote, "Protecting privacy in the computer age is like trying to change a tire on a moving car."

We can summarize:

Instead of merely protecting personal data, an organization must ask whether any new data including aggregate statistics, can be used in combination with data from other sources to deduce information about individuals. In addition, other aspects of security must be chosen to prevent violations of privacy.

17.11 Back Doors, Side Channels, And Other Concerns

Interestingly, privacy considerations intertwine with other aspects of data security. For example, one attack on privacy starts by targeting cloud-based email accounts to discover ways to steal credentials. Once credentials have been stolen, the attacker uses them to launch attacks against web applications. In 2020, Threatpost reported that stolen credentials were used in twenty-one percent of attacks involving the hacking of web applications, resulting in the exposure of more than sixty million records containing personal data.

Because multiple tenants and multiple applications from a given tenant share the underlying infrastructure, an organization must be careful to avoid possible *back doors* and *side channels* that allow data to leak to unintended recipients. Three points of contact between applications cause concern: shared storage systems, shared computing systems, and shared networks.

As we have seen, cloud systems rely on virtualization to isolate computation, communication, and storage systems. The hardware and software technologies used for virtualization work well, and most achieve complete isolation. However, instances have occurred where flaws, including flaws in the underlying hardware, allow data to transfer from one application to another through a side channel. In addition, attackers try to insert back doors into systems that send copies of data to the attacker. Although tenants cannot predict or prevent flaws from occurring, they can remain aware of the potential danger, and work hard to detect back doors.

17.12 Cloud Providers As Partners For Security And Privacy

When it moves its computing to the cloud, an organization must learn to use the configuration and management interfaces the provider offers. Unfortunately, a new interface can be difficult to master, and small mistakes can lead to security problems. According to the *2019 Data Breach Investigation Report* by Verizon, for example, misconfiguration of storage systems accounted for twenty-one percent of unwanted data exposures.

How can a tenant ensure that its systems remain secure? One answer comes from a change in overall philosophy. In a traditional IT infrastructure, the IT staff follows a maxim that keeps attackers at bay: "trust no outsiders." In a cloud environment, however, an IT staff becomes dependent on the cloud provider. Therefore, the staff must learn to view the provider as a partner. The tenant specifies its security policies and trusts the provider to help implement the policies.

The challenges that arise from learning and using a provider's configuration and management interfaces become especially complex in a multicloud environment. More important, security systems available from one provider may not be available from another. Thus, it becomes essential to work with each provider to ensure uniform policies.

A final aspect of cooperation with a provider arises when an incident occurs or when an anomaly has been detected. In some cases, a tenant may be able to analyze the problem, discover the cause, and affect a repair. In other cases, the problem may be rooted in the underlying infrastructure where only the provider has the authority needed to diagnose the cause. In the worst case, a problem may involve multiple tenants. The point is:

> *A tenant must view a provider as a partner and work closely with the provider on security, both to ensure policies have been configured correctly and to handle problems that arise in the underlying infrastructure.*

17.13 Summary

In terms of security, the cloud environment poses special concerns for a tenant. The tenant has less control and visibility into the underlying systems, making it difficult to diagnose problems. Because a tenant must share the infrastructure with outsiders, potential attackers have much closer proximity to the tenant's systems. Because cloud replicates microservices to handle scale, a tenant faces a dynamic execution environment with bursts of activity; dependencies among microservices further complicate security. The remote user access mandated by cloud means a tenant must ensure confidentiality of business data while allowing employees to communicate over the Internet. Finally, a tenant must depend on software from the cloud provider and third parties.

Many of the security techniques used in a traditional IT infrastructure do not work well in the cloud. Traditional security divides users into insiders who work for the organization (employees and contractors) and outsiders who have no relationship to the organization. In a cloud environment where potential attackers may have access to shared networks, using the insider/outsider distinction does not provide sufficient security.

As an extension of the insider/outsider distinction, traditional IT infrastructure used perimeter security, with the organization forming a conceptual fortress with security systems used to control access at external connections. As one special case of perimeter control, traditional IT infrastructure often uses a demilitarized zone (DMZ) at each connection to the Internet.

In a cloud environment, the lack of a clear border between an organization and the rest of the data center means perimeter security does not apply; tenants must adopt a perimeterless (i.e., borderless) approach that uses zero trust security. As part of zero trust, a tenant uses an identity management system that controls each user's access rights, allowing fine-grained restrictions. A Privileged Access Management system handles administrative or superuser privileges, and logs all accesses.

Artificial Intelligence technologies can be used both to break security and to strengthen security. Attackers use AI schemes to probe vulnerabilities. Organizations use AI technologies, especially machine learning, to detect anomalous behavior of users, applications, and network traffic. AI technologies can also use the context in which accesses occur to identify anomalies.

Encryption and Virtual Private Network technologies can keep data confidential when an employee uses remote access. In addition, an organization must take steps to prevent outsiders from accessing data on a user's device. An organization can use workflow security, which means a security policy stays with data, even if the data moves to an employee's device.

The availability of many information sources makes privacy (i.e., keeping a user's personal data confidential) difficult. It may be possible to combine such things as statistical aggregates of data with other sources of information to associate facts with specific individuals.

Cloud complicates privacy by increasing the risk of back doors and side channels through which personal data can leak. Although most virtualization technologies provide the needed isolation for applications, flaws in hardware and software can create side channels.

Learning to use the provider's interfaces for configuration and management can be difficult, and configuration errors can cause security problems. When defining security in a cloud environment, a tenant must change the philosophy of not trusting any outsiders and learn to view the cloud provider as a partner who can help ensure systems follow the tenant's security policies and who can identify and repair problems in the underlying infrastructure.

Chapter Contents

18

Controlling The Complexity
Of Cloud-Native Systems

18.1 Introduction

Previous chapters describe cloud programming paradigms and a software development approach used to build cloud-native software. This chapter describes some of the complexity inherent in large concurrent and distributed systems. Complexity leads to errors and unexpected behavior, and the chapter considers approaches that have been used to help manage complexity and reduce errors.

18.2 Sources Of Complexity In Cloud Systems

Previous chapters describe many factors that contribute to the complexity of cloud systems, including:

- Myriad technologies and tools
- Layers of virtualization
- Use of third-party software
- The need for elastic scale

Although each factor does indeed add difficulty to the design of cloud software, the most significant factor that makes cloud designs complicated arises from the inherent complexity involved in distributed systems design. Software engineers building cloud systems face the same problems that plague all concurrent and distributed systems, but must contend with the problems at much larger scale and in a much more dynamic execution environment than traditional distributed computing designs. The point is:

> *Cloud-native software magnifies the problems inherent in concurrent and distributed systems.*

18.3 Inherent Complexity In Large Distributed Systems

Distributed computing systems use multiple processors attached to a computer network (e.g., the Internet) to achieve parallelism. Each processor runs an operating system that allows multiple processes to execute concurrently. Such systems exhibit well-known problems, including:

- Inconsistencies among copies of data
- Vulnerability to instance explosion
- Unexpected consequences of delay
- A potential for deadlock

Inconsistencies among copies of data. A distributed system consists of multiple programs running on separate processors. Therefore, copies of pertinent data must be sent over computer networks to each processor. Except for a few specialized applications, data items change as computation proceeds. Thus, unless a system has been designed to keep copies in sync, inconsistencies arise where the copy on one processor differs from the copy on other processors, which can produce incorrect results.

Vulnerability to instance explosion. Distributed systems often create additional processes as load increases. For example, when a request arrives at a concurrent server, the server creates a new process to handle the request. The new process may invoke services on other processors, causing them to create additional processes. Thus, a system that creates an instance for each request without any limits is vulnerable to overload when a flood of requests arrives. In a cloud environment, where each new instance typically consists of a container, the system may not reach limits quickly, but creating arbitrary instances may make the tenant vulnerable to arbitrary charges.

Unexpected consequences of delay. The modules of a distributed system communicate over a computer network, and the delay depends on the network traffic. Traffic from other systems can change delays drastically, leading to unexpected behavior.

A potential for deadlock. Most concurrent and distributed systems use synchronous communication among a set of processes. That is, a process sends a request to a service and blocks until a reply arrives. Process A can block waiting for process B, which blocks to wait for process C. Once C sends a reply, B can finish and send a reply to A, which can continue running. However, a potential exists for *deadlock* in cases where dependencies among a set of processes form a cycle. That is, each process in the set remains blocked waiting for another, with the last process waiting for the first. We will discuss such dependencies later in the chapter.

18.4 Designing A Flawless Distributed System

The inherent problems described above raise many questions about cloud software, including:

- Is it impossible for a large, complex system to be flawless?
- How will a set of interdependent microservices behave in the wild?
- Can flaws be detected before a system has been implemented?
- Can a tool help a designer know where to look for potential flaws?

It may seem that the answers lie in exhaustive testing (i.e., testing all possible computations and all possible interactions among processes). However, the possible execution sequences in a distributed system make exhaustive testing infeasible. To see why, consider communication. Recall that the delay on networks connecting processors varies. More important, one cannot merely test the system with delays ranging from short to long because delays vary over time. One would need to test the case where only the first message is delayed, only the second message is delayed, and so on. Then, one would need to test the case where the first and second messages are delayed, the second and third messages are delayed, and so on until all possible combinations for all possible delays have been tested. For a system designed to send tens of thousands of messages each minute and a range of possible delays, the time required to test all combinations that can occur as the system runs would take years. The point is:

Exhaustive testing of a large distributed system requires an infeasibly long time.

18.5 System Modeling

If designers cannot rely on testing to uncover flaws in cloud software systems, what can they do? One approach centers on modeling. The idea is straightforward: create a model of the system and use the model to predict and understand how the system will behave.

A variety of models and modeling tools have been created that can help designers understand various properties of distributed systems. The models can be divided into two broad categories:

- Operational models
- Analytical models

Operational models. An operational model uses a simulator to mimic a running system. A simulator does not perform actual computation nor does it run in real time. Instead, the simulator only focuses on key events, such as the exchange of messages among processors. The simulator estimates how the running system will behave and when each event will occur. For example, the simulator can estimate how long a given program will compute after a message arrives and where the program will send the next message. Similarly, instead of measuring the transmission of messages over a network, a simulator estimates the delay that each message will experience.

Simulation can help designers understand the network traffic patterns a distributed system will generate under various loads. However, the results of a simulation depend on how well the simulator mimics the real system. Consequently, designers face a difficult choice. On the one hand, building a simulator that represents a real system accurately can take substantial time and effort. On the other hand, a simulator built quickly may produce invalid results.

Analytical models. Analytical models help a designer understand how a distributed system will behave before the system has been implemented. How can one understand a system before code has been written? The general idea is to express each of the pieces in a high-level, abstract way, and then analyze the abstract representation to reason about the program.

Since the beginning of digital computers, programmers have used abstract program representations to reason about code and ensure correctness. For example, early programmers used flowcharts, a pictorial representation of the control flow in a program. Many programmers now use algorithms written in pseudo code as a high-level abstraction that can be analyzed before implementing the program. Although such abstractions hide the details of a programming language and make it easier for humans to understand the essence of a computation, they still require a human to imagine what will happen when a program runs. The next sections describe analytical models that allow a human to use mathematics to understand program behavior.

18.6 Mathematical Models

The question arises, "Can one create a model of software that can be used to analyze correctness without requiring a human to imagine the software running?" The Programming Language research community has considered the question for decades, and has explored several approaches. The models use various forms of mathematics, including:

- First-order logic†
- Temporal logic
- State machine models
- Graph-theoretic models

Each form defines a set of terms and symbols, along with their precise mathematical meaning that allow one to express facts about a program and then reason about the

---

†First-order logic is also known as *predicate logic*, *predicate calculus*, and *quantificational logic*.

facts. In particular, various forms allow one to reason about the program's correctness or performance. For example, temporal logic allows one to write formulas that express desired properties of *safety* (e.g., two copies of the code will not cache a data object at the same time), and *liveness* (e.g., the program will send a response when it receives a request). For now, it is not important to understand technical details or the underlying mathematics. Later sections contain examples that help explain the concepts further.

18.7 An Example Graph Model To Help Avoid Deadlock

With respect to the problem of deadlock described above, a question arises, "Can a designer ensure that a software system remains deadlock free?" One might imagine a tool that performs the needed analysis. The imagined tool could read the source code for the system and report whether the system will eventually deadlock. Unfortunately, no such tool can be built because the problem is equivalent to the well-known *halting problem*, meaning that solving the problem is mathematically impossible.

If we cannot build a tool that finds all deadlocks, what can be done? Although tools cannot identify deadlocks absolutely, a tool can be built that identifies modules that could lead to deadlock and those that could not. Such a tool has two advantages. First, in a system with hundreds of modules, the tool may identify cycles that a human would not notice. Second, the tool saves the designer time by eliminating modules that are not part of any deadlock cycles.

One way to construct a tool that can identify potential deadlocks consists of building a graph of dependencies among modules. The tool can then analyze the graph to find cycles. The designer considers each of the cycles to look for potential deadlocks. We will see additional ways a dependency graph can be used.

As a simplified example, consider a system with nine modules that invoke one another synchronously. Input to the tool consists of a list of modules along with the modules they invoke. Figure 18.1 shows an example specification.

Module	Invokes
finance	vendors, billing
vendors	records, logging
shipping	inventory, finance
orders	shipping, billing, inventory
records	logging, orders
billing	logging
logging	(none)
inventory	accounting
accounting	logging, billing

Figure 18.1 An example specification listing a set of modules and the modules each one invokes.

A tool can convert the specification into a directed graph in which each node represents a module and an edge from node *A* to node *B* means that module *A* invokes module *B*. For a large system, the graph contains many nodes and edges. For the trivial example in Figure 18.1, the graph can be displayed as a small diagram. Figure 18.2 shows one possible layout. Clearly, a graph makes it much easier to identify potential deadlock cycles than a textual list of dependencies.

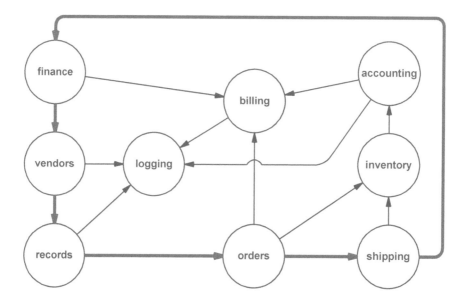

Figure 18.2 A pictorial representation of the graph for the dependencies from Figure 18.1 with a cycle shown highlighted.

18.8 A Graph Model For A Startup Sequence

As a second application of a graph model, consider starting a set of communicating services following a power failure. That is, assume that multiple services must be started. A given service may use other services during startup (e.g., a file storage service may invoke a time service to obtain the current time and set the local clock). If a service is not available when needed, the invoker will try repeatedly. If a dependency cycle exists, a form of deadlock known as *livelock* will occur in which each service repeatedly tries to invoke another service, but no responses arrive because none of the services has successfully started.

A graph model can be used to analyze startup dependencies. Once a graph has been constructed, two results arise. If a dependency cycle exists, the tool will identify the cycle as the previous section describes. If no dependency cycle exists, the graph can be used to produce a startup sequence that avoids unnecessary useless attempts to invoke a service that has not started. That is, the sequence starts modules with no dependencies first, then starts modules that depend only on those modules, and so on, until all modules have been started.

As an example, consider the startup dependencies that Figure 18.3 lists.

Module	Needs At Startup
manager	configure, logging
database	logging
time	(none)
naming	logging
domain	configure, logging
logging	time

Figure 18.3 A list of modules and the other modules that each invokes during startup.

To compute a startup sequence, one can reverse the idea of dependency by building a predecessor graph where an edge from node X to node Y means module X must be started *before* module Y can be started. Once a graph has been built, the nodes can be arranged into levels that give the startup sequence, as Figure 18.4 illustrates.

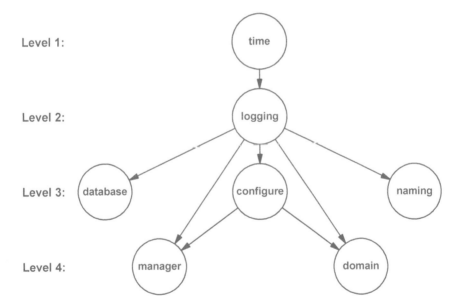

Figure 18.4 A predecessor graph for the modules in Figure 18.3 arranged into levels. Modules at level i must be started before modules at level $i+1$.

18.9 Modeling Using Mathematics

As an example of a more mathematical approach to modeling, consider TLA$^+$†, a modeling system that has been used for cloud software by companies, including Intel, Amazon, and Microsoft. Unlike the simplistic graph model described above, TLA$^+$ allows a designer to create a model (called a *specification*) that expresses complex properties of a program. The designer can then use a tool to analyze the specification and check for correctness and other conditions.

TLA$^+$ has the following characteristics:

- Is based on mathematical logic
- Models a high-level abstraction of a real system
- Uses the notion of system state and state transitions
- Includes a high-level interface for algorithms
- Is especially difficult to learn and master

Is based on mathematical logic. TLA$^+$ requires properties and conditions about a program to be expressed using mathematics. The system cannot derive a specification from source code, but instead requires a human to use mathematical notation. A TLA$^+$ specification resembles a mathematical derivation or proof more than a computer program. Instead of being integrated into the course code, a TLA$^+$ specification is kept separate. In general, a designer uses TLA$^+$ first. Once the TLA$^+$ specification has been analyzed and shown to have the desired properties, source code can be written. Thus, we think of TLA$^+$ as part of the early software design process when a designer considers the algorithms rather than as part of the implementation stage.

Of course, implementation mistakes can occur when writing the code, and TLA$^+$ cannot prevent such errors. Because it can uncover fundamental flaws, however, TLA$^+$ can prevent serious problems that are difficult for humans to identify and expensive to repair after a system has been built.

Models a high-level abstraction of a real system. TLA$^+$ describes abstract properties of a program without giving details. For example, a designer might use TLA$^+$ to specify that x is an integer without giving implementation details, such as whether x is thirty-two bits long or sixty-four bits long, whether x is represented in ones-complement or twos-complement arithmetic, or whether x is stored in big-endian or little-endian order. Similarly, a designer might specify that q and r are processes, without saying whether q and r run on the same processor or on processors connected by a network.

To understand TLA$^+$ imagine trying to write a mathematical proof-of-correctness for a computer program. One must write basic assertions, such as, 'After the assignment of 3 to variable x, the variable has the value 3." One can use the assertions to draw conclusions (also expressed mathematically). The system allows a proof to be written in a machine-readable way, and the system includes a tool that can check the proof.

---

†A reviewer noted that TLA$^+$ is a variant of TLA, which stands for *Temporal Logic of Actions*, and TLA is yet another *Three Letter Acronym* (i.e., TLA is a TLA).

Uses the notion of system state and state transitions. Rather then specifying operational steps the way a programming language does, TLA$^+$ describes the global *state* of a system by giving the values for all variables. The specification describes how variables change as the system transitions from a given state to a next state. Like most mathematical formulations, TLA$^+$ does not specify an exact outcome. Instead, it specifies the set of possible outcomes, which allows a human to reason about the possible next states of a system and their properties.

Includes a high-level interface for algorithms. In its basic form, TLA$^+$ does indeed resemble mathematics more than a programming system. To make TLA$^+$ more palatable to software engineers, TLA$^+$ includes a high-level interface that handles common situations. Named *PlusCal*, the interface allows one to enter a specification that appears more like an algorithm. A tool translates a PlusCal specification to an equivalent TLA$^+$ specification, thereby hiding many details from the user. Although most software engineers will find it less intimidating than TLA$^+$, PlusCal has limited power, meaning that it may be unable to handle all the specifications for a system.

Is especially difficult to learn and master. As with any powerful technology, TLA$^+$ comprises many constituent pieces. The system spans a broad set of possible uses. One must learn to express properties and statements about a program mathematically. Goals that seem intuitive and straightforward can be extremely difficult to express in precise mathematical terms. For example, it can be difficult to state that a system will not deadlock or a processor will be shared fairly among all processes. Consequently, one does not learn TLA$^+$ easily or quickly; mastery requires substantial effort. The point is:

> *Although it can be practical and helpful, a mathematical modeling system, such as TLA$^+$, can be intimidating and difficult to master.*

18.10 An Example TLA$^+$ Specification

The description of TLA$^+$ above gives an overview and lists its general characteristics. A concrete example will help clarify TLA$^+$ further. Of course, a small example cannot include all capabilities and complexities of TLA$^+$. However, the example will illustrate some of the mathematical notation that TLA$^+$ uses and show that even a trivial specification requires substantial effort.

Our example uses a pattern that arises in many cloud systems: a *proxy*. When an application needs to contact a service, the application sends a request to a proxy, which forwards the request to one of the target instances of the service. When the target responds, the proxy forwards the response to the application. Each request contains a sequence number that the application uses to match a response to a request. The target returns the sequence number that it receives along with its response. Figure 18.5 illustrates the flow of a message from an application through a proxy to a target instance and back.

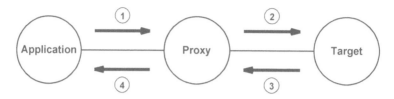

1: An Application sends a request to a Proxy
2: The Proxy forwards the request to a Target
3: The Target sends a response to the Proxy
4: The Proxy forwards the response to the Application

Figure 18.5 The four basic steps taken when an application sends a request
to a target through a proxy.

18.11 System State And State Changes

A TLA$^+$ specification describes the total state of the system. That is, a specification gives the values of all variables. In practice, mathematical models do not attempt to represent all data items associated with each process. Instead, a mathematical model abstracts away details and focuses only on a few key items. Our example focuses on the messages being sent as key values to be considered. Thus, our TLA$^+$ model only needs to specify variables associated with messages, and does not need to specify the variables used by the application, proxy, and target processes. In more complex cases, of course, the state of a system will contain many variables.

The state of a system (i.e., the values of variables) changes over time. Thus, in addition to specifying a set of variables, TLA$^+$ must also specify how the variables change. In the proxy communication example, when a target responds to a request, the target sends a new response. Conceptually, the system state contains a response variable, and the value of the variable changes when a target responds. In a programming language, one writes an assignment statement that updates a variable. For example, the following adds 1 to variable x:

$$x = x + 1$$

In mathematics, the above equation makes no sense. To denote a modified value, mathematicians append an apostrophe to a variable name and pronounce it *prime*. Therefore, to specify that the value of variable x increases by 1, a mathematician writes:

$$x' = x + 1$$

The equation makes sense mathematically because it is not self-contradictory. TLA$^+$ employs the mathematical approach. Thus, when reading a TLA$^+$ specification, one must remember that an equation involving the prime version of a name represents a modification of a variable rather than a new variable.

Interestingly, TLA$^+$ does not have an explicit representation of time. Instead, one must use items in the state to specify temporal sequences. To understand how time can be represented by state variables, consider the four numbered steps in Figure 18.5. To model the steps, one might choose to define a variable named *step*. When the step variable has value 1, the values of other variables represent the state of the system during step 1 in the figure. Similarly, when the step variable has the value 2, the value of other variables represent the state of the system during step 2, and so on.

18.12 The Form Of A TLA$^+$ Specification

A specification begins with a module name and declarations analogous to those found in a computer program. Figure 18.6 lists some of the keywords used in declarations.

Keyword	Meaning
EXTENDS	Reference to pre-defined sets
CONSTANTS	Declares names of constants
VARIABLES	Declares names of variables

Figure 18.6 Example keywords used in declarations in a TLA$^+$ specification.

The *EXTENDS* keyword allows a specification to reference pre-defined sets. For example, TLA$^+$ defines the set *Naturals* to be the natural numbers (i.e., non-negative integers), and the set *Integers*, to be the set of all integers. It also defines *TLC* to be a set of utility functions used by the model checker. Thus, in addition to other items, a specification usually EXTENDS TLC.

Because it includes a notion of time sequences, TLA$^+$ allows one to specify items for which the value changes over time as well as items that have a fixed value. The keyword *CONSTANTS* declares the names of items that retain a single, unchanging value, and the keyword *VARIABLES* declares the names of items for which the value can change over time.

TLA$^+$ does not specify a single sequence of state transitions. Instead, a TLA$^+$ specification lists conditions under which each transition can occur. Conditions can overlap, which means that more than one transition can occur at a given time. The idea is that a TLA$^+$ specification covers all possible orders of transitions.

To describe state transitions in TLA$^+$, the second section of a specification contains a set of *state predicates*. Each predicate contains a list of conditions that are required to be true to make activation of the predicate possible (i.e., to make the transition possible). When reading a specification, one must remember that predicates can be activated in any order, and the specification includes all possible orderings. An example will help clarify the idea. Figure 18.7 contains a TLA$^+$ model for proxy communication with three principal predicates named *Application*, *Proxy*, and *Target*. A predicate named *Init* gives initial conditions.

```
┌─────────────────────── MODULE proxy ───────────────────────┐
 EXTENDS Naturals,TLC
 CONSTANTS Request, Response, NULL, Rounds
 ASSUME NULL ∉ Request
 VARIABLES  step, req, resp, seq

 TypeInvariant  ≜ ∧  resp ∈ [ Response → Response ∪ { NULL } ]
                  ∧  step ∈ { 0, 1, 2, 3, 4 }
                  ∧  req ∈ [ Request → Request ∪ { NULL } ]
├─────────────────────────────────────────────────────────────┤

 Init    ≜          ∧  step = 0
                    ∧  seq = 0
                    ∧  req = NULL
                    ∧  resp = NULL

 Application ≜       ∧  ∨  ∧  step = 0
                          ∧  step' = 1
                          ∧  req' ∈ Request
                          ∧  seq ' = seq + 1
                          ∧  seq < Rounds
                       ∨  ∧  step = 4
                          ∧  step' = 0
                          ∧  UNCHANGED <<req, seq>>
                    ∧  UNCHANGED <<resp>>

 Proxy   ≜          ∧  ∨  ∧  step = 1
                          ∧  step' = 2
                       ∨  ∧  step = 3
                          ∧  step' = 4
                    ∧  UNCHANGED <<req, resp, seq>>

 Target  ≜          ∧ step = 2
                    ∧ step' = 3
                    ∧ resp' ∈ Response
                    ∧  UNCHANGED <<req, seq>>

 Next  ≜            Application ∨ Proxy ∨ Target

 Spec  ≜            Init ∧ □ [Next]_<<step, req, resp, seq>>
├─────────────────────────────────────────────────────────────┤

 THEOREM Spec ⟹ □ TypeInvariant
└─────────────────────────────────────────────────────────────┘
```

Figure 18.7 A TLA$^+$ model for proxy communication.

18.13 Symbols In A TLA$^+$ Specification

To understand predicates in the specification, one must know the symbols used and their meaning. TLA$^+$ defines two forms for each symbol: the form used in mathematics publications and an ASCII form that can be used to create a machine-readable specification. Figure 18.8 lists some of the TLA$^+$ symbols along with the equivalent ASCII form and meaning of each.

Symbol	ASCII	Meaning
⌐‾‾‾‾⌐	- - - - - - - - -	Start of module (with MODULE name embedded)
├─────┤	- - - - - - - -	Section separator in a module
⌊‾‾‾‾⌋	- - - - - - - - -	End of module
∧	/\	Logical and
∨	\/	Logical or
∪	\union	Union of two sets
∩	\intersect	Intersection of two sets
→	->	[S → T] means f(x) such that x∈S implies f(x)∈T
≠	#	Not equal
∈	\in	x ∈ S means x is an element of set S
∉	\notin	x ∉ S means x is not an element of set S
≜	==	x ≜ y means x is defined to equal y
⇒	=>	a ⇒ b means a being true Implies b is true
□	[]	□ F means F always remains true

Figure 18.8 Symbols used in a TLA$^+$ specification along with the equivalent
ASCII representation and the meaning of each symbol.

As an example, consider the *Target* predicate that specifies how a target sends a response when it receives a request.

$$
Target \triangleq \quad
\begin{aligned}
&\wedge \quad step = 2 \\
&\wedge \quad step' = 3 \\
&\wedge \quad resp' \in Response \\
&\wedge \quad UNCHANGED \;\; <\!<req, seq>\!>
\end{aligned}
$$

Each predicate specifies conditions that must be met before the predicate can be activated. Most important, the first condition of the *Target* predicate specifies that the *step* variable must have the value 2. The second condition specifies that the value of variable *step* will become 3. The third condition specifies that the *resp* variable will take on a new value drawn from the *Response* set (i.e., when it receives a request, the target sends a response). Finally, the fourth condition specifies that the values of variables *req* and *seq* will remain unchanged (i.e., the target returns a copy of the request along with the response).

The *Proxy* predicate handles two cases. For step 1, the proxy forwards a request to a target, and for step 3, a proxy forwards a response back to the application. Thus, if the value of variable *step* is 1, it becomes 2, and if the value is 3, it becomes 4. The values of other variables remain unchanged. To express the two cases, TLA$^+$ uses a *logical or*. In the specification, indentation has been used to show that the two cases are part of the first condition.

$$
\text{Proxy} \triangleq \quad \wedge \ \vee \ \wedge \ \text{step} = 1 \\
\wedge \ \text{step'} = 2 \\
\vee \ \wedge \ \text{step} = 3 \\
\wedge \ \text{step'} = 4 \\
\wedge \ \text{UNCHANGED} \ \text{<<req, resp, seq>>}
$$

The *Application* predicate also contains two cases. When *step* is zero, the application can send a request, and when step is 4, the application receives a response. When sending a request, *step* becomes 1, the *req* variable takes on a value from the *request* set, and the sequence variable *seq* is incremented by 1. A note below explains the condition that ensures *seq* remains less than *Rounds*. When receiving a response, *step* is reset to 0, and the *seq* remains unchanged, making it possible for the *Application* predicate to be activated for another round.

$$
\text{Application} \triangleq \quad \wedge \ \vee \ \wedge \ \text{step} = 0 \\
\wedge \ \text{step'} = 1 \\
\wedge \ \text{req'} \in \text{Request} \\
\wedge \ \text{seq'} = \text{seq} + 1 \\
\wedge \ \text{seq} < \text{Rounds} \\
\vee \ \wedge \ \text{step} = 4 \\
\wedge \ \text{step'} = 0 \\
\wedge \ \text{UNCHANGED} \ \text{<<req, seq>>} \\
\wedge \ \text{UNCHANGED} \ \text{<<resp>>}
$$

The *Next* predicate forms a key part of the specification. It uses *logical or* to combine the *Application*, *Proxy*, and *Target* predicates. In essence, *Next* specifies that any of three can occur: the application can send a request, the proxy can forward the request or the response, or the target can receive a request and send a response. Thus, in the mathematical model, the *Next* predicate specifies that the system components act asynchronously.

$$
\text{Next} \triangleq \quad \text{Application} \vee \text{Proxy} \vee \text{Target}
$$

In practice, conditions control when a predicate can run, and the conditions have been designed to restrict the sequence of predicates. For example, the *Target* predicate cannot become active until the *step* variable has the value 2. The only way that *step* will become 2 is when the *Proxy* becomes active. Similarly, the *Proxy* cannot become active until the *step* variable has the value 1, and the *step* can only become 1 once the *Application* predicate becomes active.

The *Spec* predicate gives the overall specification, which consists of one activation of *Init* followed by an arbitrary number of activations of *Next*. In other words, *Spec* uses the "always" temporal operator to specify what a programmer thinks of as indefinite iteration.

$$\text{Spec} \;\triangleq\; \text{Init} \wedge \square\,[\text{Next}]\_{<<\text{step, req, resp}>>}$$

The final item in the specification consists of a mathematical theorem that states the *Spec* predicate implies that the *TypeInvariant* always remains true.

$$\text{THEOREM Spec} \Rightarrow \square\,\text{TypeInvariant}$$

A model checker can read the specification and check that the theorem does indeed hold. In fact, the THEOREM specifies an arbitrary number of activations of *Spec*. A model checker cannot handle an arbitrary number of iterations. Therefore, the example specification uses constant *Rounds* to allow one to limit the number of messages that can be sent before the model checker stops. When running the model checker, a programmer must assign a value for *Rounds* according to how long they are willing to wait for the model checker to run and how much memory is available for the model checker to use. Each time the checker activates the *Application* predicate with *step* equal to zero, the sequence number variable, *seq*, will be incremented. Once the sequence value reaches *Rounds*, the *Application* predicate becomes false. Once the *Application* predicate stops setting *step* to 1, no other predicates can become active, and the model checker stops.

18.14 State Transitions For The Example

To understand how the predicates specify state transitions, consider a short sequence of transitions. The example state contains only four variables as specified in the model:

$$\text{VARIABLES}\quad \text{step, req, resp, seq}$$

The state of the system consists of a value for each variable. The model checker displays state transitions by showing a sequence of values for the variables. The display allows one to understand transitions possible according to the specification. Instead of merely displaying values, the model checker displays a mathematical expression for the state. Figure 18.9, which follows the approach used by the model checker, shows a small transition diagram for one round of a message passing from an application through a proxy to a target and back. To keep the diagram small, *req* has the value 1, *resp* has the value 3, and *seq* stops after one round.

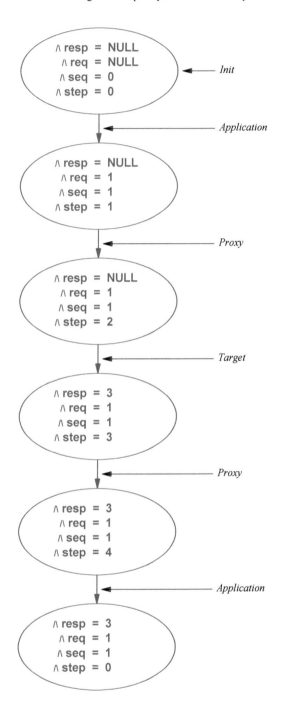

Figure 18.9 State transitions for one round of sending a message and receiving a reply. The label on each transition tells the predicate that caused the transition.

18.15 Conclusions About Temporal Logic Models

Three facts about mathematical models that use temporal logic should now be apparent. First, the abstraction that a mathematical model describes differs dramatically from actual software. Second, formulating a model requires one to understand mathematical logic as well as the details of a particular modeling technology. Third, it may be difficult to create software that follows a model.

One uses a model to help conquer the complexity of system design. Ironically, a model can introduce additional complexity. For example, writing a correct TLA$^+$ specification can be extremely difficult, even if one understands the general idea. Programmers accustomed to thinking about specifying algorithmic steps and iterative processing must change to a mindset that uses a mathematical interpretation in which equations and logic expressions describe the set of all possible outcomes. More important, details matter because a minor mistake in a specification can have major consequences for correctness†.

We conclude:

> *Although a mathematical model may help a software engineer understand properties of a system, using such a model introduces additional complexity because it requires an engineer to learn to think in new ways, master concepts from mathematical logic, and learn the details of a language to express a model as well as the tools used to check its validity.*

18.16 Summary

Complexity in cloud computing arises from many factors, including myriad technologies and tools, layers of virtualization, third-party software, and scale. A significant aspect of complexity arises from the problems inherent in concurrent and distributed systems.

Because it would take an infeasibly long time to test a large distributed system, one cannot rely on exhaustive testing to find flaws. Instead, a designer must build a model of the system and then use the model to reason about the system operation. An operational model uses a simulator to mimic events in a system. An analytical model uses a mathematical abstraction to allow one to reason about a system.

Mathematics used for models has included first-order logic, temporal logic, state machine models, and graph-theoretic models. Although it cannot absolutely identify whether a system will deadlock, a graph-theoretic model can be used to show a designer the circular dependencies that may lead to deadlock. A graph-theoretic model can also be used to identify a startup sequence that ensures a module does not attempt to start until all modules on which it depends have already started.

---

†The author, who holds an undergrad degree in mathematics, spent hours learning the basics of TLA$^+$ and helping a colleague create the example. Misunderstandings about the syntax and semantics as well as small errors in entering a specification led to many unexpected problems.

As an example of mathematical modeling, we examined TLA$^+$, which uses a form of temporal logic. TLA$^+$ models a system by describing the state of all variables and defining possible transitions from one state to the next. A TLA$^+$ specification uses mathematical notation. TLA$^+$ defines an ASCII equivalent for each symbol, which allows a specification to be read and analyzed by a model checking program. Proponents of modeling argue that building a model can help overcome the complexity inherent in a complex system and make it easier to understand its properties. Mathematical models introduce additional complexity by requiring one to understand the fundamentals of mathematical logic and learn the details of a specific modeling language and associated tools.

Index

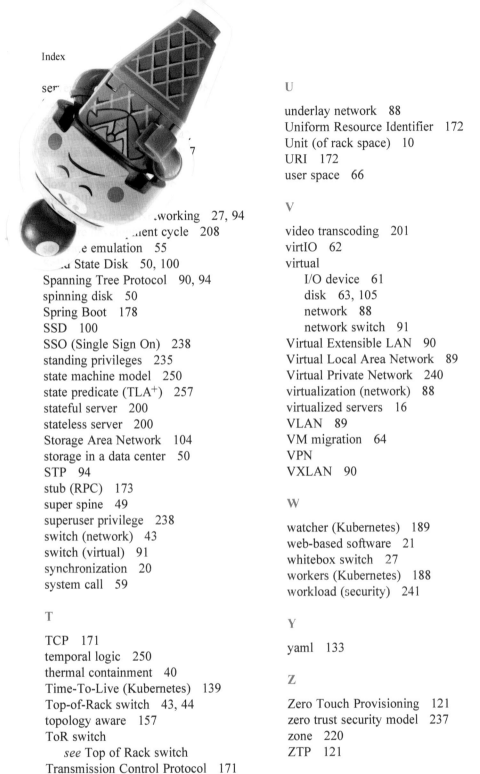

Milton Keynes UK
Ingram Content Group UK Ltd.
UKHW050450071024
449327UK00014B/316